Data Analysis with SPSS
A First Course in Applied Statistics

Third Edition

Stephen A. Sweet
Ithaca College

Karen Grace-Martin
Cornell University

Boston New York San Francisco
Mexico City Montreal Toronto London Madrid Munich Paris
Hong Kong Singapore Tokyo Cape Town Sydney

ISBN-13 978-0-205-48387-7
ISBN-10 0-205-48387-9

Printed in the United States of America

10 9 8 7 6 5 4 3 2 1 11 10 09 08 07

Contents

Preface

Much has changed since we published the first edition of this book over a decade ago. Students have become more adept at using software, and their colleges and universities have increased their capacities to bring computer technologies into the classroom, and more hands-on learning experiences are being integrated into college courses (McKinney, Howery, Strand, Kain, and Berheide 2004; Sweet and Strand 2006). But also, the volume and complexity of data used in public discourse has continued to expand, which in turn continues to raise the bar for the skills needed to navigate an increasingly complex world (Association of American Colleges & Universities 2002).

This book fits into a larger mission to cultivating quantitative literacy (Howery and Rodriguez 2006; Madison and Steen 2003; Shaeffer 2003; Sweet and Strand 2006). While we have composed data sets that are of considerable interest to students in sociology (but also economics, psychology, demography, family studies, and other social sciences), they can be used by statistics teachers in other disciplines as a means to teach the essentials of statistical methods. We have written less about statistics than about how to take real world observations, expose these observations to statistical analysis, and understand how to systematically study social relationships. We provide an active hands-on approach to learning the skills of data analysis, using the Statistical Package for the Social Sciences (SPSS) and some of the most current data on social behavior available.

We believe that statistics are best learned in *context,* as they are applied to analyzing real world observations. Our hopes are that readers will realize what we learned long ago, that statistical analysis (contrary to popular belief) can be fun. This assertion will strike many readers as surprising and counterintuitive. We think one of the reasons traditional statistics courses are not well received by most students simply stems from focusing too much on the formulas that *generate* statistics, and not enough on how statistics are *interpreted.* In contrast to the methods of calculating statistics, the interpretation of statistics involves applying the appropriate statistical methods and then analyzing the results to gain an understanding of people and society.

We also believe that statistics become interesting when they are *applied.* We do not view statistics as a means of testing mathematical skills; rather, our interests concern how statistics can be used to gain understandings of the social world. Do social policies work the way they were intended? Does a social program make a difference? In what ways do different groups of people vary from one another? In this book, we try to chart effective pathways to answering these types of research questions using real data in the same manner that professional social scientists would.

To develop skills, we introduce key concepts and statistical approaches commonly used by social scientists, and then conclude chapters with hands-on exercises. These exercises draw from two data sets, rich with information concerning social behavior. One of these data sets contains 161 variables that document social behavior and public policy in the United States, focusing on differences between states. These variables are drawn from the 2000 Census and other sources, offering opportunities to study a variety of topics including education, criminality, welfare use, health care, taxation, births, mortality, and environmental well-being. The other data set contains nearly 133 variables from the 2004 General Social Survey conducted by the Roper Center for Public Opinion. These data enable analysis of variation in individual experiences and attitudes relating to education, religion, politics, sexuality, race, and taxation. These data sets are used to build analytic skills in a step by step manner, from the initial stages of forming research questions and collecting data through advanced multivariate methods and report writing. As these two large data sets offer abundant information on places and people, they open a vista of possibilities for independent research activities.

Do I Have What It Takes and What Will I Accomplish?

There is something about statistics that seems sparks fear. This is unfortunate and we believe unnecessary. Some readers (possibly you?) may be concerned in their mathematical skills. This text is designed to help you become refined in analyzing quantitative data; therefore existing skills in statistics and math can certainly be a help. However, this book is designed to teach basic statistical inference. Therefore, no prior statistics courses and only modest mathematical capabilities are needed to perform the exercises in this text.

Finally, there is the issue of relevance. Some readers may wonder why they should bother learning a statistical program that will unlikely have application in their intended profession. The short answer is that this book is not so much about SPSS, as it is about the methods to create and interpret statistics. To our eye, SPSS is to the statistician what the hammer is to a carpenter – it is simply a means of getting a job done. All of us live in a world in which numbers are used to explain our experiences, and can be used to inform, or mis-inform. Understanding the ways numbers used can help us position ourselves in respect to important issues, ranging from the causes of cancer, global warming and welfare reform (to name but a few hotly contested issues). Understanding how data are analyzed can help all readers develop a critical perspective on how to evaluate arguments based upon quantitative data. Indeed, the real aim of this book is not so much about teaching SPSS. It is to cultivate the skills needed to be a critical consumer of information and to whet the appetite for empirical inquiry.

Acknowledgments

Many people contributed to this book at various stages of its development. Scott Morgan and Kathleen O'Leary Morgan were exceedingly generous in allowing us to reproduce the data presented in their outstanding books: *State Rankings 2007, Health State Rankings 2007,* and *Crime State Rankings 2007*. The Roper Center for Public Opinion Research allowed us to select and include variables from the 2004 *General Social Surveys*, data from one of the very best public opinion polls available.

We are indebted to Jim Rothenberg and Susanne Morgan for their careful reviews of this edition. We also thank Kathy Wood for her work in proofreading the manuscript. Jeff Lasser, our editor, at Allyn and Bacon, offered excellent advice throughout the project.

Most of all, we thank Karen's husband Michael, and Stephen's wife Jai, who gave unwavering love and support.

About the Authors

Stephen Sweet's interest in social relationships led him to pursue a doctoral degree at the University of New Hampshire. There he found himself more than a bit surprised that his statistics courses engaged his imagination on various ways to understand people and society. And in the process of becoming a social scientist, he also developed a passion for teaching students not only about social processes, but also the skills needed to apply sociological perspectives to their lives and communities. He has published extensively on issues relating to work and family lives, including the books *Changing Contours of Work* (Pine Forge Press 2008), *Work and Family*

Handbook: Interdisciplinary Perspectives, Methods and Approaches (Lawrence Erlbaum Associates 2005), and *College and Society: An Introduction to the Sociological Imagination* (Allyn and Bacon 2001). Currently, Stephen Sweet teaches in the sociology department at Ithaca College and works with Sloan Work and Family Research Network to develop teaching resources for work-family scholars and students.

Karen Grace-Martin's years as a social psychology researcher at UC Santa Barbara and the University of Kent, when she was bewildered by the many statistical methods available, led her to pursue a degree in applied statistics. Since then, she has excelled in explaining statistics in non-technical ways to both students and applied researchers. Karen was a statistical consultant at Cornell University for seven years and has taught statistics courses for economics, psychology, and sociology majors at Santa Barbara City College and the University of California, Santa Barbara. Currently, Karen does freelance statistical consulting, workshops, and training, and can often be found in her pottery studio or organic garden with her two homeschooled children.

Dedication

With admiration, we dedicate this book to our students, who continually impress us with their questions, observations, creativity, and perseverance.

Chapter 1
Key Concepts in Social Science Research

Overview

We begin our introduction to data analysis with the major concepts social scientists use to develop research questions, understand data, and test relationships. This orientation to statistical reasoning and research provides a foundation for using SPSS as practiced in professional social science. Issues include the reliance of social science on observation and reason, consideration of the ways causal relationships are established, the formation of testable hypotheses, and the identification of valid and reliable measures. All of these concerns are relevant to analyzing the data provided in this book and successfully launching an independent research project.

Why Do We Need Statistics?

Social scientists spend their professional lives studying two overarching sets of questions: (1) how do social arrangements affect the human condition, and (2) how do humans influence these social arrangements? This dual nature of relations between individuals and wider society leads to a number of interesting questions. For example, how does growing up in a poor neighborhood affect an individual's life? Will it influence the prospect of graduating high school? Will it increase the likelihood of being victimized by crime? Does it contribute to a sense of fear? What other effects might it have? Alternately, one can invert the direction of the relationship between the individual and society to consider the ways people influence their environments. For example, if people have strong feelings of patriotism, does it affect their tendency to vote? Will these values influence their inclination to vote for Republican, Democrat, or Independent candidates? Do political allegiances vary by age or by birth cohort? What other personal qualities affect their political behavior?

It is possible to speculate on the nature of any number of social relationships, but ultimately resolving these questions requires the collection and analysis of data. This is the unique contribution of the social sciences (economics, sociology, psychology, political science, anthropology), as these professions employ scholars who rigorously apply logic and observation to the study of social arrangements. It is this push for empirically based understanding that has driven the need for statistics. Reliance on statistical methods emerged because scientists agreed that it was not enough to make conclusions based on a few observations, abstract thoughts, or taken-for-granted beliefs.

Consider the controversial issue of the death penalty. Some argue that stiff sanctions (such as the death penalty) deter crime, while others argue that not only do they not deter crime, they are disproportionately applied to minorities. These beliefs can be tested statistically by comparing the crime or homicide rates in places that have the death penalty with those that do not[1] and by comparing the ratios of racial minorities and Whites receiving the death penalty. It is important to emphasize that statistics will not resolve debates about the morality of policies such as the death penalty, but they are a powerful tool to test assumptions on the impact of social policy on behavior and life chances.

Mark Twain once quipped: "There are lies, damn lies, and statistics."[2] His skepticism of the cavalier use of numerical facts and figures is as relevant today as it was a century ago. All too often bits and pieces of data and relationships are selectively reported with the intent to sway people rather than inform them. And, contrary to the old maxim, numbers do not speak for themselves. Sometimes the biggest challenge is in determining what numbers to pay attention to! In this book, we hope to cultivate the skills that will enable readers to offer objective and nuanced interpretations of observations, which might otherwise be colored by emotions, political motives, or even ignorance.

Those who are most skilled at data analysis approach their research with a commitment to **value neutrality**. Their goal is not to prove a particular finding, but to explore data, in all its complexity, and present meaningful findings. They will have to decide the types of information to collect and use, how to rework data so that it becomes amenable to analysis, and the types of analytic techniques to apply. And even if it produces uncomfortable results, their primary commitment is to offer an objective reading of what the data indicate, to fully report findings, and to present understandable reports on what are oftentimes complicated sets of relationships. As empiricists, they set aside their politics and resist the tendency to "massage" data to only present findings that support their beliefs. But this is not to say that skilled data analysts are anything less than passionate. Their passion is in understanding the complexity of society and how different facets of culture and structure intersect with the ways people chart their lives. There is no problem with political, religious, and other civic-oriented passions driving the types of questions posed, but they should not determine the answers!

Framing Topics into Research Questions

Many novice researchers begin their projects by saying "I want to show that _____" (insert whatever thesis you wish). They then expend considerable effort to collect pieces of information to sustain their preexisting perspectives. But this is not the scientific approach. Social scientists begin studies with questions, opening up the possibility that the evidence at hand may not support their initial beliefs. In reality, one would be hard pressed to find scientists who do not want to see their theories supported; yet it is by opening assumptions to critical analysis that they advance knowledge. For most projects, there comes a point at which researchers, on the basis of

---

[1] Are you curious about this issue? The data included in this book will enable you to test this relationship.
[2] Autobiography of Mark Twain (2000 [1917]).

whatever data they have, come to conclusions. These conclusions may support their theses, they may refute them, or they may indicate that additional data will be needed to resolve their questions. The research question should drive social science inquiry, not the desire to establish a particular belief as being "true."

The first step in any research project is to identify the **research question**. Developing a research question involves considering an issue and posing it in the form of a problem that can be answered. This will put the researcher on a pathway that sets clear goals, establishes the required types of data, and structures the collection and analysis of information. Ideally, research questions fill voids in the knowledge base, and seek answers to problems that have not been sufficiently addressed. In this manner, research questions create intriguing puzzles, and their resolution generates a sense of fulfillment. Social scientists pose questions so that unexpected results can emerge and be seriously considered. "Is there racial discrimination in the education system?" is a much better way to pose a research project than "I want to show that there is discrimination in the education system." The problem-posing approach helps researchers create a clear analytic strategy. For example, some interesting sub-questions about discrimination in education are:

> Is there an equitable representation of minority groups within the curriculum?
>
> Is there proportionate employment of minority members in school systems?
>
> Do graduation rates vary among racial groups?
>
> Do minority school districts receive comparable funding to those in predominantly white school districts?

Notice that these questions require different types of information. The first question suggests looking at what is taught in classes, the second focuses on employment patterns, the third on graduation rates, and the fourth on school funding. In developing research questions, novice researchers (as well as seasoned scientists) often realize that their initial ideas are too broad to be managed in one study. When this happens, a good strategy is to parse the project into manageable sub-questions, and select those questions that should be immediately pursued. It is also necessary to consider issues of **feasibility**—are there data available that will address the question? What types of resources would it take to generate usable data? In this book are two data sets, one with data at the individual level (the study of people), and another with data at the societal level (enabling the study and comparison of states). Our hopes are that you will be able to use these data sets to generate interesting and feasible questions to gain immediate research and statistical experience.

The work of social scientists rests on the shoulders of those who preceded them. For this reason, as you work toward developing a study, it is important to perform a **literature review** in the early phases of the project. The literature review is an overview of past research on topics relating to the research question. The literature review does three things. First, it helps the researcher appreciate the knowledge that has been gained on a subject to date. Second, it informs the researcher of the methods and analyses other scholars used to answer similar research questions. Finally, it helps the researcher contribute to the accumulation of knowledge by integrating current research with previous literature. When researchers examine the literature and identify still unanswered questions or yet untried methods, they position themselves to perform new and vital work.

The internet has greatly increased the accessibility to perspectives and articles related to research questions you may want to pose. But not all sources are created equal. The reality is that anyone with a computer and an internet link can post information on the web, and there are no checks on the veracity of what they argue. For this reason, the best sources for literature reviews are scholarly journals such as *The American Journal of Sociology*, *Child Development, Journal of Marriage and the Family*, and *Journal of Health and Social Behavior*, to name a few. The articles in these types of journals have undergone a rigorous review procedure that insures (as best as one can hope) that what is presented is verifiable and accurately presented. Most articles in scholarly journals have a detailed description of how data were collected and offer extensive references to other related studies. These sources are available through key word searches of computer databases located in college and university libraries.

Theories and Hypotheses

Finding answers to research questions involves developing theories and testing hypotheses. A **theory** is an explanation that connects observations and reasoned analysis into an integrated meaning system. Sometimes the word "theory" is used to disparage a position and to suggest that an explanation (such as global warming or evolution) is only one idea among many other potentially equally valid ideas. But scientists embrace the concept of theory in a different way. For them, a theory is something to be both respected and systematically challenged. The respect for a theory comes from its ability to demonstrate internal consistency and withstand alternate explanations. For example, a theory of global warming would have a hard time withstanding stable temperature patterns, and Darwin's theory of evolution would not be favored if animals could will themselves to grow tails.

But theories are also subject to challenge, and central to the scientific method is the need to test alternate explanations and to see if a given theory can be applied to uncharted areas. For example, a theory of racial bias may be well established in the workplace, but what about in the classroom? The wider its application, the stronger its predictive capabilities, and the more it withstands any tests that could dislodge it, the stronger a theory becomes. But it is important to note that no theory can be absolutely proven, as there exists a possibility that a better explanation can be found. For example, under the Ptolemaic system, there was considerable evidence to suggest that the earth was the center of the universe, as the sun, the planets, and the stars all appeared (with some irregularities) to revolve around the earth. But when Copernicus placed the sun at the center of the solar system, many of the irregularities disappeared, such as the perception of Mars occasionally sliding backward in its orbit.[3]

Data analysts oftentimes assume the role of a puzzle solvers, taking theories and dismantling them, testing assumptions, and observing the extent to which data fit predictions. In other words, given what a theory would suggest, do the data operate in the predicted manner? To make these predictions, researchers generate **hypotheses**—testable statements concerning expected sets of relationships. For example, one theory is that racial discrimination remains a pervasive feature of American society. If this is the case, one can make a series of predictions

---

[3] As Thomas Kuhn observed in *The Copernican Revolution* (1997), even the old "earth centered" model could accommodate many of these irregularities in planetary movement. Mars, for instance, was thought to orbit in a pattern that approximated a "figure 8," with the earth at the center of the lower loop. The critical question for data analysts sometimes is not whether a theory can work, but is that explanation better than a competing theory.

about the ways discrimination will be manifested – in schools, in the workplace, in religious organizations, etc.

The data analyst oftentimes works in a deductive manner, taking a theory and dismantling it into component parts, constructing hypotheses, and then exposing the hypotheses to tests. At the end of the process the analyst will give an appraisal of the theory and the extent to which the data supported its assumptions, as well as the amount of variation in the data that remain unexplained (did the theory explain much of what is observed in the data, or only a little?). But this is not the only approach to analyzing data. In other circumstances, researchers operate more like Sherlock Holmes and inductively deduce what the data have to offer. This **grounded theory** approach relies on exploring the data, posing more tentative hypotheses, and building a theory out of these observations. This is then applied to other data sets and, in that manner, builds a theory up from observation, but then cycles that theory back into further analysis. In other words, data analysis is oftentimes not as linear as it seems, as one is frequently observing, testing, and exploring relationships to be found in the data.

Population and Samples

As researchers set up their hypotheses, they should consider the **population**, the entire group of observations they are interested in studying. The conversational use of the term population usually means all people in a country. This could be the population in a research study, but so could all elementary-aged children in a city, all college applicants in a region under a certain admissions policy, all nursing home residents who receive Medicaid in a state, etc. Also, it is not only people that can comprise a population, as populations can be the number of businesses or schools in a society, states in a nation, or nations in the world.

Although the population makes up the total number of elements (persons, states, schools, businesses, states, countries) that are of interest in a research study, it is rarely possible to measure every single element in a population. Even if you could get the names of every elementary-aged child in the country, the resources necessary to study all children would be astronomical. And even if it were possible to get the needed funds and their contact information, some potential participants would "age out" of the study before they could be observed. And since each year there are always new elementary-aged children, the study would have to go on forever. Because of the many practical reasons for not measuring the entire population, engaging in social science most often involves studying a **sample**, a subsection of the population. The sample is a smaller set of elements, pulled from a larger population.

What makes a sample well chosen? Usually, the primary consideration is that the sample has the characteristics of the population under consideration, and, if it does, it is considered a **representative sample**.[4] For example, a representative sample of people in the United States would be roughly half male and half female, roughly one in ten would be African American, and roughly one in five would be under the age of fifteen. These ratios would remain the same if the sample was composed of fifty, one hundred, or one thousand individuals. But there are instances when a researcher may want to over sample specific subpopulations, and in those instances various "weights" can be introduced to adjust the analysis for the impact these discrepancies introduce between the sample characteristics and the population characteristics.

[4] But there are also instances when researchers will want to generate non-representative samples as well, such as in case studies of single organizations or events.

Relationships and Causality

Most social science research is concerned with **causality**, determining how particular sets of conditions lead to predictable outcomes. Seldom, though, are relationships **deterministic**, producing inevitable results. More often, social science documents **probabilistic relationships**, in which factors increase or decrease tendencies toward particular outcomes. For example, consider the conclusion that education increases income. If this were a deterministic relationship, everyone with a college degree would have higher incomes than everyone with a high school degree. This, of course, is not the case. But it *is* the case that those with college degrees *tend* to earn more than those with high school degrees. Thus causality does not mean an inevitable outcome, only that one factor will tend to push another factor in a predictable direction.

How does a researcher determine if a relationship is causal? Answering this question requires identifying the relationships between independent and dependent variables. **Independent variables** are hypothesized to cause changes in **dependent variables**. We have already discussed a number of hypothesized causal relationships, such as the death penalty deterring crime and school funding increasing graduation rates. Diagrams of these relationships would be:

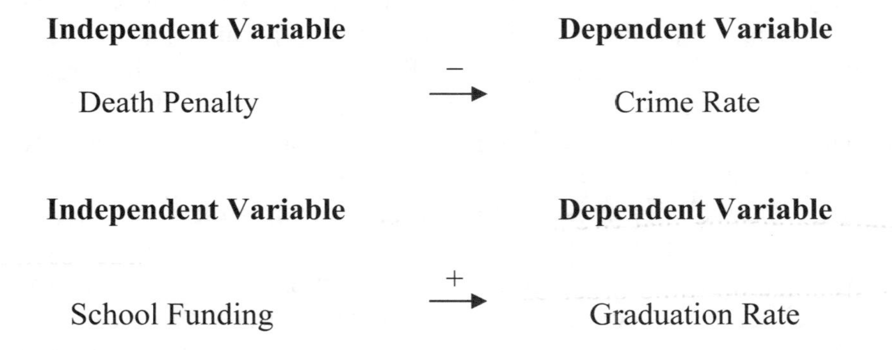

The very same indicators can switch from being independent variables to dependent variables, depending upon the research question being addressed. For example, increased educational success could potentially lead to lower crime rates:

Independent Variable **Dependent Variable**

Graduation Rate $\xrightarrow{-}$ Crime Rate

To assert a causal relationship is to claim that changes in the independent variable create changes in the dependent variable. In practice, researchers should only assert that one factor causes changes in another variable when they can reasonably satisfy the following three criteria:

1. *Association*: there must be a relationship between the independent and dependent variables.
2. *Time order*: the change in the independent variable must precede the change in the dependent variable.
3. *Nonspuriousness*: the effect of a third unmeasured "spurious" factor does not produce the relationship between two variables.

Association

When two variables are associated, the variables correspond with each other in predictable ways. **Associations** can either be **positive** (an increase in one variable corresponds with an increase in the other) or **negative** (an increase in one variable corresponds with a decrease in the other variable). For instance, the statement "people with more education tend to have higher incomes than people with less education" is a statement of positive association. "People with more education tend to be less prejudiced than people with less education" is a statement of negative association. Note that neither statement claims a causal relationship, only that changes in the variables correspond with one another.

It is important to note that many associations are not causal. For instance, red cars are in more accidents than white cars. Red cars do not cause the accidents, it is the tendency of aggressive drivers to purchase these red cars and get into accidents (see the discussion of spurious relationships below).

Time Order

When a change in one variable causes the change in another variable, logically, the independent variable needs to change *before* the dependent variable. To invert time order leads to irrational conclusions. For example, does it make sense to conclude that a mousetrap sprang because the mouse died, or that it rained because the ground got wet? But consider this dynamic. Pornography is oftentimes considered a cause of violence against women. Let us put time into the equation. If there is a relationship, one would expect that a substantial rise in pornography use would be *subsequently* followed by a rise in violence. Has the increasing consumption of pornography over the internet contributed to this dynamic? Thus, identifying causality requires going beyond establishing that two variables are associated with one another, but also that one variable *leads* to a change in the other variable. The issue of time order seems straightforward enough, yet, establishing time order can be difficult because most data are cross sectional and reflect observations collected at one point in time. As such, asserting the existence of a causal relationship has to be done with care. Also, as time often introduces changes in other variables, even when time order is satisfied, causality may still not be the case (see below).

Nonspuriousness

A **spurious relationship** exists when two variables appear to be causally related, but their relationship owes to the influence of a third unidentified variable. In spurious relationships, if the third unmeasured variable is taken into account, the relationship between the initial two variables disappears. A classic example of a spurious relationship is that of ice cream consumption causing drowning deaths. There is an association between ice cream consumption and drowning (more people drown during times when a lot of ice cream is being consumed). The time order issue can also be satisfied (increases in ice cream sales precede increases in drowning deaths). Of course the unmeasured factor in this relationship is temperature. More people swim in the summer and more ice cream is consumed in the summer. It is in fact the warm weather operating as a spurious factor, causing both variables to increase.

Spuriousness is difficult to rule out simply because it is impossible to account for every variable that could produce the appearance of a relationship. In actual practice, good research involves measuring the factors that could potentially produce a spurious relationship. For example, even though sunspots *might* have an effect on the behavior a researcher is studying, there is no theoretical reason for this assumption, so there is no need to account for it in the analysis. In contrast, income and education may have a profound impact on many aspects of social life, so introducing these factors in statistical analyses is generally a good idea.

In sum—unless data analysts can reasonably satisfy the issues of associations, time order, and nonspuriousness, they should avoid asserting that a relationship is causal. Causality may, in fact, exist, but the data analyst should only go as far as the data allow.

Data Sets

This book introduces social science research by applying statistical methods to two data sets—one documenting the characteristics of people, the other documenting characteristics of places. We depart from traditional methods for teaching statistics by focusing less on statistical calculations and more on interpreting what statistics reveal. We try to create a context where newcomers can apply rigorous statistical methods to their own empirical studies. This research-oriented approach teaches the skills needed for applied data analysis. Toward this end, the data sets included with this book are structured to enable newcomers to immediately put statistics to practical use.

The **General Social Survey (GSS04) data** contains information on people. These data are based on interviews of men and women who responded to a series of questions concerning their attitudes, behaviors, and beliefs. Questions include, among other things, demographic information such as ethnicity, income, age, and education. Other questions tap into beliefs and behaviors, such as the degree to which individuals hold prejudicial attitudes and the amount of television they watch. As you work through the exercises in this book, and consider your own research projects, it may help to think of the GSS04 data as being about people. Who are these individuals? What are their beliefs? What are their characteristics? How do they behave? As we discuss later, some of the most interesting questions involve comparing people with one another, on the basis of race, gender, socioeconomic status, education, and other socially constructed divisions.

The other data set, the **STATES07 data**, contains information on geographic socio-political areas. This information includes poverty rates, alcohol sales, crime rates, and a variety of other information about life within each state. The sources of these data include the U.S. Census, National Education Association, Department of Labor, the Internal Revenue Service, and the U.S. Department of Health and Human Services. These are among the most reliable sources of information on the social condition of the nation. The STATES07 data offer information on the contexts in which people's lives are lived. As such, the concern for most of the exercises and projects that involve these data is on collective behavior and social operations.

Parts of a Data Set

This book addresses how to approach a data set in a systematic manner and how to launch a research project involving statistical analysis. Any research project requires collecting and understanding **data**, pieces of organized information about elements in a population (a **datum** is a single piece of information). We have taken care of the first task and collected some of the best quality data available to date on people and places. In fact, these two data sets are comparable, and in many instances identical, to the types of data that professional social scientists use to study social life. They are also the types of data policy makers use to inform legislative decisions. Our hopes are that these data will also help you understand statistics and how data analysis is applied in social science research. Those already beginning to think about research projects may find it useful to skim the appendices at the end of this book, where the data sets are described in greater detail, to see if these can provide information to address existing research interests. For others, this information may help kindle ideas and spark the imagination for a wide variety of potential projects.

Data sets are structured to link cases with variables. A **case** is the individual unit being studied. Each case has one datum, or **observation**, for each variable. For the GSS04 data, each case is an individual person. In the STATES07 data, each case is the individual state, such as New Hampshire. **Variables** contain information about each case and are structured in a way that makes one case comparable to another. For instance, in the GSS04 data, all of the respondents were asked their age, and, as a result, one could compare any one respondent's age with that of another. The STATES07 data have variables indicating crime rates, and therefore states can be ranked in terms of how much crime occurs within their geographic areas. When data are collected systematically, researchers can analyze the patterns between variables. For instance, having information on crime rates and whether the death penalty exists in the fifty states allows a researcher to analyze the relationship between these variables. Although the two data sets included in this book can help expedite many research projects, there are many other sources of data available over the internet. As researchers consider using these resources, they need to consider the strengths and weaknesses of these resources, as data quality can vary markedly.

Reliability and Validity

When selecting variables for study, the scientist strives to find **indicators** that best capture abstract concepts. For example, studies of criminality require using some indication of the occurrence of crime. One possibility is to use data from surveys assessing victimization, asking people if they have ever been crime victims. Because many crimes are never reported to the police, this indicator will likely offer different information than arrest records or police reports. Studies of educational achievement could involve examining grade point averages, graduation rates, or scores on standardized tests.

As researchers select and construct indicators, they strive to make them as reliable and valid as possible. **Reliability** refers to the degree to which an indicator is a consistent measuring device. It is no accident that carpenters use metal tape measures instead of strings and navigators use compasses instead of star gazing. Strings and stars work, but not nearly with the consistency offered by the other methods. The same concern extends to social research, and researchers strive to select consistent tools for measuring concepts such as alienation, alcohol consumption, and depression. Consider the implications of assessing educational achievement by asking students "how well do you do in school?" Unless all students use the same standards for self-assessment, the results for different students will reflect their own perceptions, not their achievements. Because reliability of such measures is questionable, SAT scores, or other standardized test scores are more favored measures. But even these measures are subject to criticism. For this reason, the very best studies use multiple indicators to examine the same underlying concept.

While SAT scores and other standardized tests offer consistency, they are often criticized for failing to accurately measure what they purport to measure. For example, SAT (standardized aptitude tests) supposedly document students' abilities to learn. But in reality, they are rather weak in their ability to predict which students will, or will not, be successful in college. The concept of **validity** addresses this concern, as it considers the extent to which the indicator measuring what it is supposed to appraise. No one uses a steel ruler to measure temperature, simply because it was never designed for that purpose. Although this concern seems straightforward, novice researchers sometimes select indicators that are loosely related to the abstract concept they are interested in studying. For example, IQ scores are sometimes erroneously used to measure educational achievement. Poverty is inappropriately used as a measure of criminality. The error in both examples is confusing a preconceived relationship with the variables intended to be measured.

Summary

To recap, social science is built on empiricism, and for many studies this requires the use of statistics. Statistics measure patterns of individual variables and relationships between them. As social scientists develop studies, they set aside their personal agendas in favor of a reasoned analysis of data, reporting results accurately and without bias. As they develop their research projects, they select or collect data that offer the most accurate indicators for the concepts they are interested in studying, seeking variables that are reliable (consistent) and valid (appropriate).

Social science research hinges on the careful construction of research questions, the pursuit of the highest quality data, and the best analytic strategies. It requires researchers to think critically about their beliefs and assumptions about the ways the social world operates. This involves creating theories and distilling them into testable hypotheses. When hypotheses are constructed, data need to be gathered and organized in a manner that will enable statistical tests. Even after statistical analyses are performed, researchers have to reflect on their theory and methodology and assess the degree to which any observed relationships can be attributed to causal processes or to other potential explanations.

Key Terms

Association
Case
Causal relationship
Data
Datum
Data set
Dependent variable
Deterministic relationship
Feasibility
Grounded theory
Hypothesis
Independent variable
Indicators
Literature review
Negative association
Observation
Population
Positive association
Probabilistic relationship
Reliability
Representative sample
Research question
Sample
Spurious relationship
Theory
Time order
Validity
Value neutral

Chapter 1 Exercises

Name JOSH Date ____________

1. Identify the independent and dependent variables in the following research projects:

A. A study seeks to find out if listening to heavy metal music causes teenagers to become more violent than their peers who do not listen to heavy metal music.

Listening to Metal — Independent Variable

violence — Dependent Variable

B. A group of researchers is interested in examining the effects of long-term poverty. They do this by studying subjects' physiological health and attachment to the workforce.

long term poverty — Independent Variable(s)

Health/ attachment workforce — Dependent Variable(s)

C. A study finds that self-esteem increases as a consequence of receiving good grades on examinations.

Good grade — Independent Variable

selfestem — Dependent Variable

2. Indicators are used to measure abstract concepts. List some indicators that might prove to be reliable and valid in measuring the following concepts. Note that sometimes social indicators can be drawn from observing people, but as we show in our examples, they can also be drawn by considering distributions of resources, organizations, or other sources of data.

A. Economic Prosperity — E.g., Gross National Product

____

____

____

B. Family Violence — E.g., Admissions to Battered Women's Shelters

____

____

____

C. Consumer Cultures — E.g., New York Times Bestsellers List

____

____

____

D. Mental Health — E.g., Survey Reports of Happiness

____

____

____

3. Using Appendix 1, identify the variable labels and variable names from the STATES07 data that might be good indicators for the following concepts:

A. Educational Attainment

EDS128 — Variable Name   High school grad rate — Variable Label

EDS147 — Variable Name   Higher ed rate — Variable Label

EDS73 — Variable Name   HS drop out rate — Variable Label

B. Health

HTH98 — Variable Name   Infant Mort. Rate — Variable Label

HTH386 — Variable Name   STD Rate — Variable Label

HTH501 — Variable Name   % Adults who don't exercise — Variable Label

D. Poverty

~~EDS48~~ PVS493 — Variable Name   Poverty Rate: 2005 — Variable Label

PVS495 — Variable Name   % children in Poverty — Variable Label

PVS527 — Variable Name   % on Food Stamps — Variable Label

4. Using Appendix 2, identify the variable labels and variable names from the GSS04 data that might be good indicators for the following concepts:

A. Religious Commitment

ATTENDA — How often R attends Religious Service

Variable Name — Variable Label

B. Prejudice

AFFRMACT — Favor Affirm. Action

Variable Name — Variable Label

WORKBLKS — Blacks hard working/lazy

Variable Name — Variable Label

C. Sexual Activity

HOMOSEX — Homosexual sex relation

Variable Name — Variable Label

XMARSEX — Sex w/ person other than spouse

Variable Name — Variable Label

D. Family Structure

RESPNUM — # in Family

Variable Name — Variable Label

ADULTS (23) — Household members 18 yrs. or older

Variable Name — Variable Label

5. Researchers have found a consistent relationship between schools and crime. More crimes occur in neighborhoods surrounding high schools and junior high schools than in neighborhoods far away from schools. The researchers conclude that schools cause crime and believe that this relationship may have something to do with teachers not teaching students the appropriate lessons in the classroom. On the basis of these data, do you find this argument compelling? Why or why not? If not, what could explain this relationship?

No. Schools themselves are inanimate. A more likely explanation is that crimes are committed by younger people and because there are usually more young people around schools, there is more crime.

6. The U.S. Department of Justice has found that of the children killed by their parents, 55% of the murders were performed by the child's mother and 45% of the murders were performed by the child's father. A researcher uses these data to support his contention that women are more violent than men. On the basis of these data, do you find this argument compelling? Why or why not? If not, what could explain this relationship?

No. This is such a broad generalization taken from a very specific finding. It could be that b/c women traditionally spend more time w/ children, they are more likely to be in positions where they might kill them.

7. A social commentator argues that the welfare programs which were introduced in the mid 1960s have caused an unparalleled expansion of poverty in the United States. Based on the following data from the *Statistical Abstract of the United States 2007*, would you agree or disagree with this causal statement? Explain.

Year	Percent Below Poverty Level
1960	22%
1965	15%
1970	13%
1975	12%
1980	13%
1985	14%
1990	14%
1995	14%
2000	13%
2005	13%

I would disagree. obviously the % has dropped or stayed the same since the 1960s.
The only leg the researcher could stand on is that the gross # has increased.

8. What environmental conditions influence crime?

A. List a few hypotheses, predictions of factors that influence crime.
E.g. Ho_1: Crime is caused by a lack of jobs

Ho_1: " " age

Ho_2: " " poverty

Ho_3: " " % of unemployed men

Ho_4: " " lack of education

B. Use the STATES07 descriptions in Appendix 1 and list variables that might help test your hypotheses.

DMS454
Variable Name

Variable Label

PVS493
Variable Name

Variable Label

ECS93
Variable Name

Variable Label

EDS130
Variable Name

Variable Label

C. While examining the Appendix 1, did you find any additional variables that may also influence crime?

NO

______________ ____

Variable Name Variable Label

______________ ____

Variable Name Variable Label

______________ ____

Variable Name Variable Label

______________ ____

Variable Name Variable Label

9. What characteristics of individuals are associated with happiness?

A. List a few hypotheses, predictions that influence happiness.
E.g. Ho_1: The wealthier a person is, the more happy they are.

Ho_1: ______________________________

Ho_2: ______________________________

Ho_3: ______________________________

Ho_4: ______________________________

B. Use the GSS04 descriptions in Appendix 2 and list variables that might help test your hypotheses.

______________	______________________________
Variable Name	Variable Label
______________	______________________________
Variable Name	Variable Label
______________	______________________________
Variable Name	Variable Label
______________	______________________________
Variable Name	Variable Label

C. While examining the Appendix 2, did you find any additional variables that may also influence happiness?

________________ Variable Name   ____ Variable Label

________________ Variable Name   ____ Variable Label

________________ Variable Name   ____ Variable Label

________________ Variable Name   ____ Variable Label

Chapter 2
Getting Started: Accessing, Examining, and Saving Data

Overview

This chapter offers an introduction to SPSS and when you complete the exercises in this chapter, you will know how to access the data sets included with this text. We also discuss some useful strategies for making data sets accessible, including naming, labeling, and defining variables.

We suggest that you work through this and the remaining chapters in this book with the SPSS program running on your computer. As we describe the commands, try to reproduce them. Feel free to explore, as mistakes are never fatal (assuming you remember to save and back up your files—always a good practice!). The chapter will conclude with some guided exercises that review some of the operations outlined here.

The Layout of SPSS

SPSS operates by presenting users with four different windows, one that presents the data, another that presents information about the data structures, another the provides the output of analyses, and another that offers a means of creating programs called "syntax." When you are in the **Data View Window,** your screen should look like the first panel in Figure 2.1. This screen looks and operates like spreadsheet programs such as Excel. Because there are no data in the Data View window when SPSS starts, the grid is empty and the Data View Window says "Untitled." But once you have data in the program, the *Data View* will display the values for each observation. There is a tab at the bottom of the window that enables you to move to a **Variable View Window**. This window will display information about each variable in the data set. Again, because we have not yet opened or defined a data set, this window is also empty.

SPSS for Windows is menu driven and is designed to fulfill common requests easily. There are two types of menus at the top of the window. The **menu bar** contains organized paths to commonly requested procedures, such as opening files, performing statistical operations, and constructing graphs. As we work through the SPSS program, you will learn many of the commands in the menu bar. You can explore some of these by drawing the cursor to a command category, such as *File*, illustrated in Figure 2.2. Within each action, SPSS will ask for the specific information it needs to perform that command. This usually includes designating the variables for the analysis and the types of output to report.

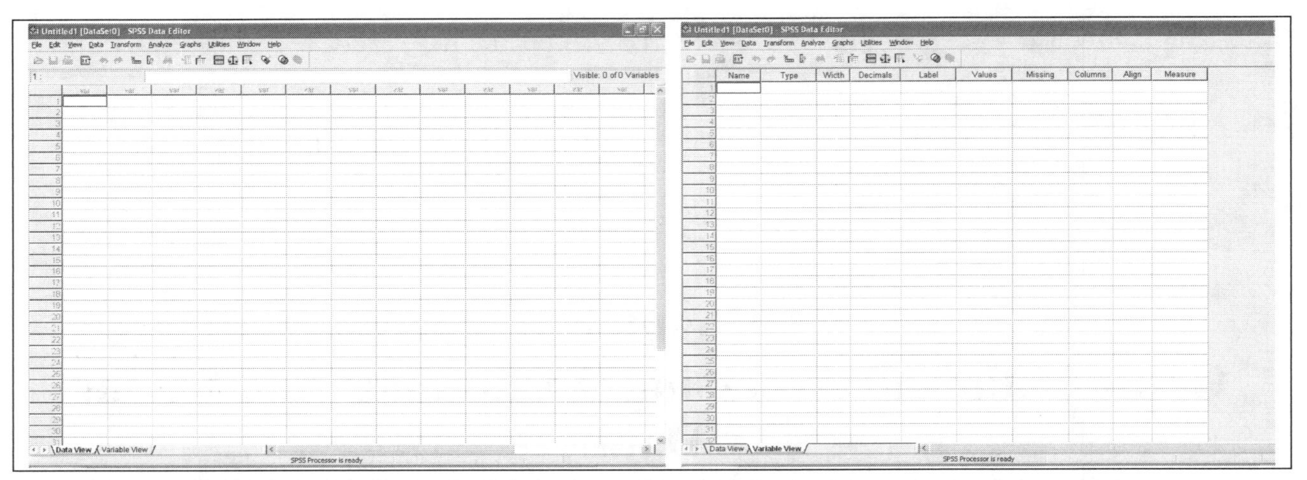

Figure 2.1 The Data View and Variable View Windows

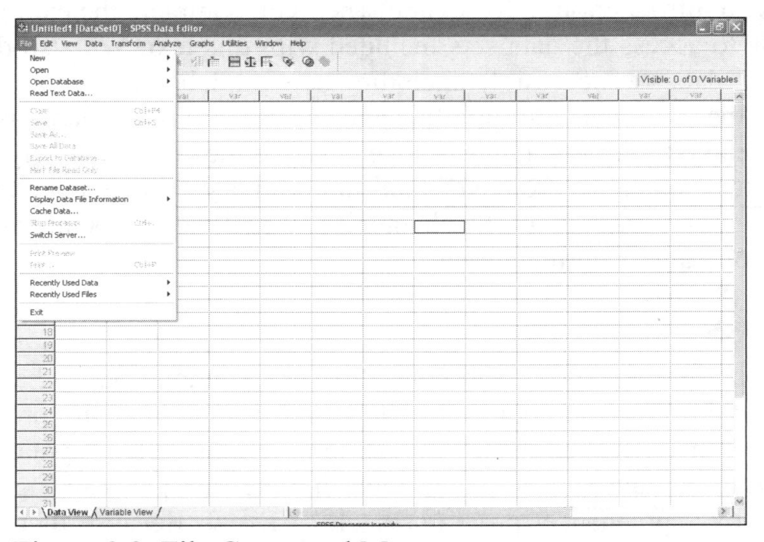

Figure 2.2 File Command Menu

In this book, we use a shorthand method to describe how to locate specific commands. For instance, if we want you to open a new data file, you will see the following:

File

New

Data

This is a quick way of saying: go to the *File* command in the main menu, hold down the mouse button, move the cursor to highlight *New,* and then move the cursor to highlight *Data*. Upon releasing the mouse button, you will have told SPSS to open a new data file and SPSS will give you a window in which you can enter the pertinent information.

Beneath the menu bar is the **tool bar**. This works like the menu and it includes a number of handy devices that refine the analyses. We will not worry about using the tool bar, but as you become more familiar with the program, you will probably find yourself curious about the types of things it can do. Feel free to experiment.

Not incidentally, there is another way to run commands in SPSS called **syntax commands**. Unlike the menu driven procedures, syntax commands require familiarity with the particular terminology and phrasing SPSS uses to perform statistical procedures. It is likely that your professor first learned to use SPSS on earlier versions of the program, which relied exclusively on these syntax commands. For complex and repetitive procedures, this is still the most efficient way to work with SPSS (in fact, this was how we constructed the data sets included in your text). However, for the purposes of the current activities, you will be served well by concentrating solely on the menu commands. We will introduce syntax later, as it can be helpful in advanced operations.

Types of Variables

Constructing a data set requires entering data (either manually or in file batches) into the SPSS program. The data will be arrayed on a grid in the *Data View* so that cases are positioned in rows, and variables are positioned in columns. Recall from Chapter 1 that **cases** are the units being studied, such as people, organizations, states, or countries. **Variables** are the characteristics or values being measured for these cases, such as age and gender for people, or per capita income and mortality rates for states.

String Variables

SPSS also expects you to designate the format of each variable. By default it will assume that you are entering numeric information. But some information may take the form of written text called **string variables** (meaning "strings" of text), dates, currencies, or other formats. But in most instances, one wants to have variables retain their numeric form, as that greatly facilitates statistical analysis. For example, we could create a string variable for gender, typing in the words "male" or "female" for each observation. But a better approach is to designate women as "0" and men as "1" and then to associate the word "male" and "female" with a value label. We will show you how to do this later.

Categorical Variables

SPSS also enables you to designate the measurement system used to construct the variable. There are several measurement levels, each reflecting different ways of assessing and documenting information about each case. One useful way of thinking about measurement is to reflect on whether variables indicate categories (such as gender or ethnicity) or numbers (such as age or income). These two types of variables, categorical and scale, often require different types of statistical analyses.

Categorical variables indicate typologies into which a case might fall. For example, ethnicity (e.g., White, Black, Hispanic, Asian) is a categorical variable. Categorical variables can be further distinguished as being either ordinal or nominal. Categories are considered **ordinal** if they can be sequenced in a logical order. For example, a student's class year in school—freshman, sophomore, junior, or senior, is an ordinal categorical variable. Knowing that one student is a senior and another is a sophomore tells us not only that the two students are in different class years, but also that the senior has completed more classes than the sophomore. Categorical variables are considered **nominal** if they cannot be sequenced in a logical order. A student's major is an example of a nominal categorical variable. It categorizes the student and gives information about what a student is studying, but there is no reason why psychology, sociology, history, or biology majors should precede or follow one another. Ethnicity is another example of a nominal variable.

Scale Variables

Scale variables (also called **numerical variables**) give information indicating the quantity of a concept being measured. For example, the number of years a person has spent in school is a scale variable, as is the number of miles a car can go on a gallon of gas. There are two types of scale variables: count and continuous. **Count variables** indicate "how many" and take on the values of whole numbers. For example, the number of classes a student is taking is a count variable. A student can take 3 or 4 classes, but cannot take 3.7 classes.

Continuous variables indicate "how much" and can have any value within a given range. Continuous variables, therefore, are not restricted to whole numbers. For example, the number of years a student has been in school is a continuous variable. A student could have been in school 13.00 or 13.72 years. The possible values are restricted to a range – the minimum possible value is 0 if a student has yet to begin school and the maximum is about 100 (if a very old person was in school his or her entire life). Note, though, that any value between 0 and 100 years is possible, including decimal values. And since decimals are possible, a continuous variable can have an infinite number of possible values (13.72 is a different value than 13.721).

Initial Settings

SPSS offers a variety of options for displaying data and output. So that what you see corresponds with the presentations in the text, before beginning, we suggest that you program your settings to match ours. Start the SPSS program, go to the *Edit* menu, and choose *Options*.[5] Because we want to orient readers to the structure of data sets, in this book we specified that SPSS presents variable names in their file order, as commanded in the left panel of Figure 2.3. When SPSS presents output, we specified that it display both the variable names and variable labels (the left panel of Figure 2.3). You should note, however, that many users prefer variables to be listed in alphabetical order, and there is no problem with this, other than the variable lists being presented

[5] In some instances, when SPSS is first opened, it may ask you if you would like to open an existing data source or enter new data. Specify that you want to create a new data set.

in an order that is different than the way they are arrayed in the data set. As you work with the data, you will be able to determine which system works best for you.

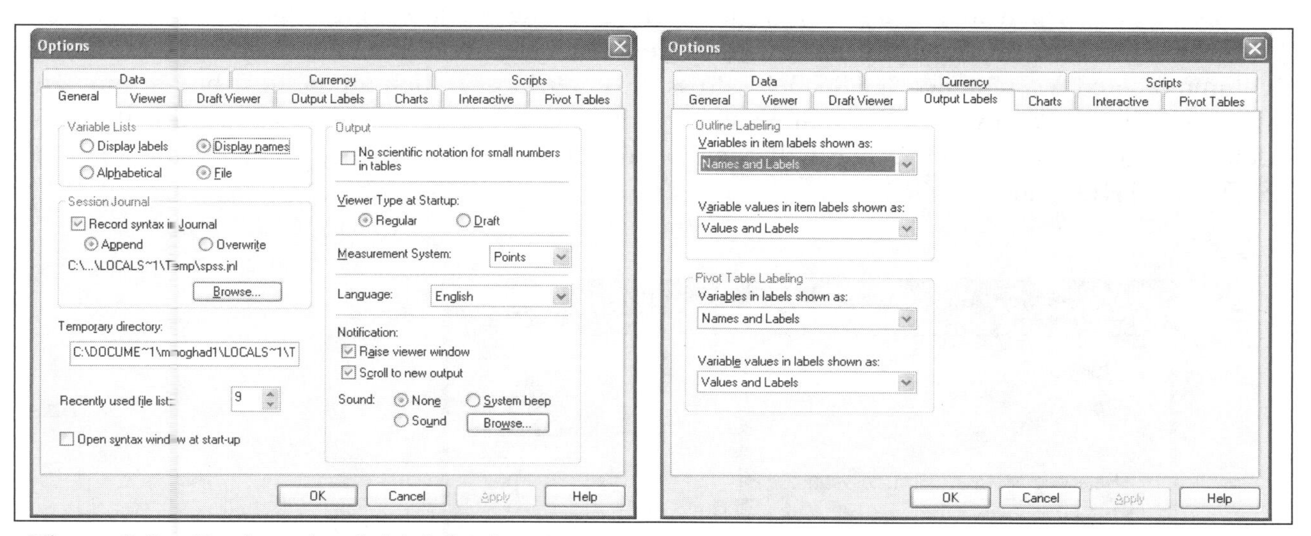

Figure 2.3 Designating Initial Settings

Defining and Saving a New Data Set

We will now begin to work with data in the SPSS environment. The first step is to load a data set into the statistical package. There are two ways to do this: manually entering a new data set or opening an already constructed data set. In this exercise, we outline how to construct a new data set. To illustrate the process, you will generate a data set about your family members. This should give you a feel for the structure of the SPSS program and how it operates. Your data set will contain four variables indicating the person's name, their sex, birth date, and level of education. Each family member will constitute one case in the data set. In the event that your family is very small, you could also include friends or other relatives (or make up a fantasy family you wish you had!). If you are working with a partner, make one dataset for the two families together.

*Untitled1 [DataSet0] - SPSS Data Editor

File Edit View Data Transform Analyze Graphs Utilities Window Help

	Name	Type	Width	Decimals	Label	Values	Missing	Columns	Align	Measure
1	PERSON	Numeric	8	2		None	None	8	Right	Scale
2	SEX	Numeric	8	2		None	None	8	Right	Scale
3	BIRTHDT	Numeric	8	2		None	None	8	Right	Scale
4	EDYRS	Numeric	8	2		None	None	8	Right	Scale
5										
6										
7										
8										

Data View / Variable View

SPSS Processor is ready

Figure 2.4 The Variable View Spreadsheet

To define the variables in your data set, enter the *Variable View Window* and replicate Figure 2.4. There are two ways to open this window. The first is to double click the cursor onto the top of any column of the *Data Editor* window. The second way (and the way to get back to the *Data View* when in *Variable View*) is to click on the *Variable View* tab in the lower left corner of the screen. Once you are in *Variable View*, you can start defining variables. In the column labeled *Name*, enter "PERSON". Press the enter key or click on the next box to move to the next column. You will notice that many boxes in the rest of the row fill in automatically with default values, some of which we will change. Add three more variables in the Name column: SEX, BIRTHDT, and EDYRS.

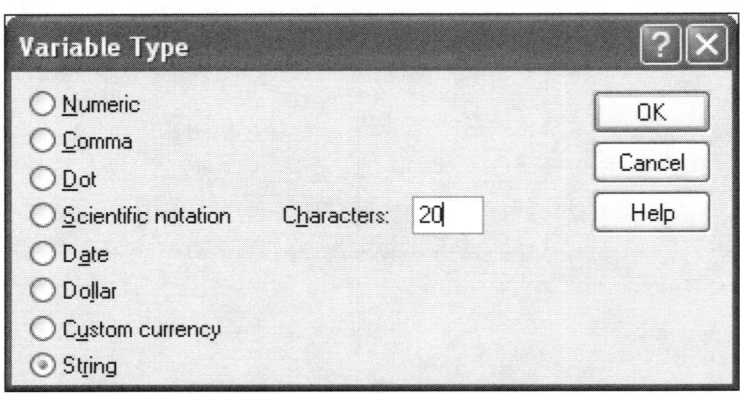

Figure 2.5 Designating Variable Types

Next we will identify the variable *Type* (designated in the second column of *Variable View*). Since we want to enter text for people's names, not numbers, we have to change the default type of the variable NAME. If we do not, SPSS will only let us enter numbers for this variable back in the *Data View* spreadsheet. To change the default, click on *Numeric* under *Type* and then click on the small gray box that appears next to *Numeric*. A window (illustrated in Figure 2.5) will pop open, in which you can choose the variable's structure. Since we want to enter text, choose *String* and enter "20" in the box for *Characters*. This informs SPSS that this variable will contain strings of non-numeric information up to 20 characters long and that the data can be treated as text rather than numbers. Click "OK" when you are done and notice that SPSS automatically put 20 in the *Width* column. At this point also designate the variable BRTHDT to be a date variable as well, with the form *mm/dd/yyyy.*

The third column *Labels,* offers the opportunity to provide a more detailed description of each variable. The following labels can be entered for each variable:

Variable Name	**Variable Label**
PERSON	Family Member Name
SEX	Sex of Family Member
BIRTHDT	Birthdate
EDYRS	Years of Education

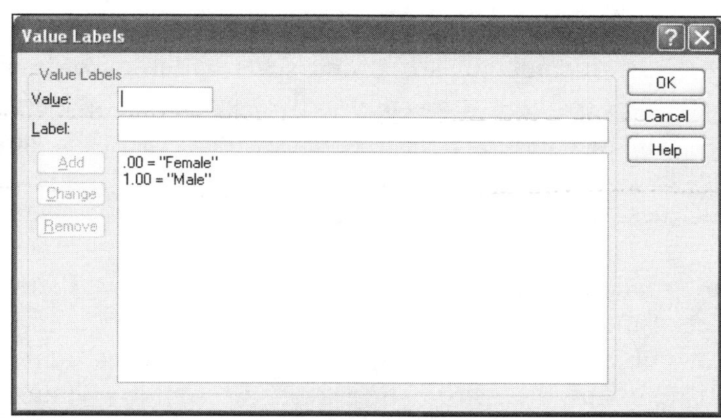

Figure 2.6 Entering Value Labels

The next column, *Values*, offers a way to associate a written description to the numerical coding of data with **value labels**. For example, we could make SEX a string variable, but we want to keep it a numeric variable. As you will see later in this book, this will enable us to engage in statistical comparisons that would not be possible if the variable was coded as string text. To do this, we will keep the variable coded as numeric and represent females with the number "0" and males with the number "1." Because it would be very easy to forget which sex we defined as "0" and "1," it is extremely important to enter value labels at this point. Click on the small gray box that appears in the cell in the *Values* column and the *Value Labels* window will pop open (Figure 2.6). Within this window, enter "0" as the *Value* and "Female" as the *Label*, and click on *Add*. Do the same for men by entering "1" as the *Value* and "Male" as the *Label*. Click on *Add*. Then click on *OK*.

For the time being, we will skip the issue of missing values, but note here the function of this column. In some circumstances researchers may want to recode certain observations so that they are excluded from analysis. For example, if a person refused to answer a survey question, or responded that they were "unsure," there may be reasons to leave those individuals out of the study. But at the same time, we would not want to lose all the other information by dropping this case from the data set. In such circumstances, missing values retain original codes, but this column designation enables SPSS to exclude them from analyses. The advantage of this system is that the decision to exclude them can be reversed by removing their designation as missing values.

*Untitled1 [DataSet0] - SPSS Data Editor

File Edit View Data Transform Analyze Graphs Utilities Window Help

	Name	Type	Width	Decimals	Label	Values	Missing	Columns	Align	Measure
1	PERSON	String	20	0	Family Membe	None	None	8	Left	Nominal
2	SEX	Numeric	8	2	Sex of Family	{.00, Female}..	None	8	Right	Nominal
3	BIRTHDT	Date	10	0	Birthdate	None	None	10	Right	Scale
4	EDYRS	Numeric	8	2	Years of Educ	None	None	8	Right	Scale
5										
6										
7										
8										

Data View | Variable View

SPSS Processor is ready

Figure 2.7 The Completed Variable View Window

Next we want to designate the measurement structure in the final column to the right. Recall that there are three distinct data structures, scale variables, ordinal variables, and nominal variables. String variables, by default, are nominal variables, as they are each unique descriptors. SEX (male vs female) is also nominal, as there is no hierarchy in the value of one sex over the other. But birthdates and years in school are both scale variables, as they are continuous variables. Designate these variables accordingly and your *Variable View* should look like Figure 2.7

SPSS is now ready for entering data. Switch to the *Data View Window* (click the tab in the lower left corner of the screen) and enter the names of members of your family. You can then enter their sex (remember to use the numeric code), birthdates, and years in school. In some circumstances, you may lack the information needed to enter a value for a particular case. For example, if you did not know your father's birth date, you should code that datum as "missing." Do this by entering a period (.) as the value. Your screen should look like Figure 2.8 when done (but of course with different information entered).

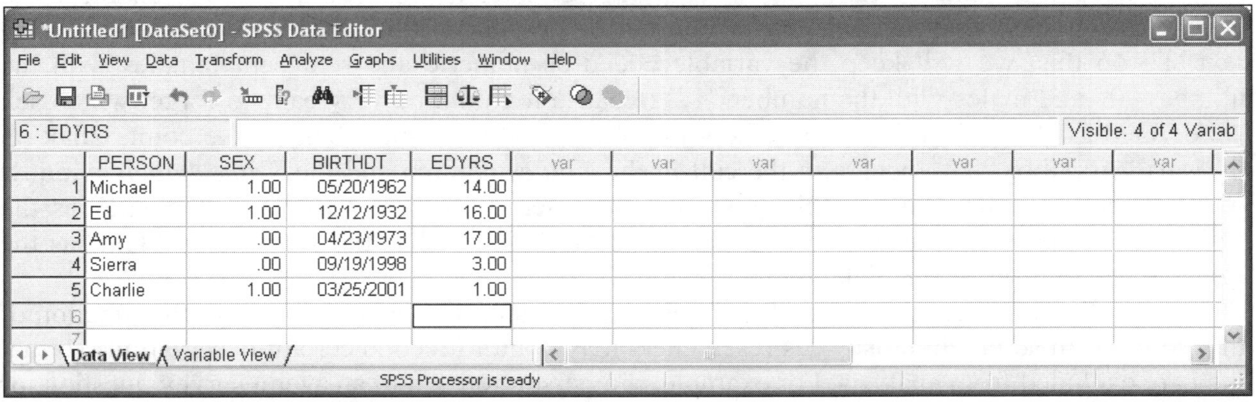

*Untitled1 [DataSet0] - SPSS Data Editor

File Edit View Data Transform Analyze Graphs Utilities Window Help

6 : EDYRS — Visible: 4 of 4 Variab

	PERSON	SEX	BIRTHDT	EDYRS	var	var	var	var	var	var	var
1	Michael	1.00	05/20/1962	14.00							
2	Ed	1.00	12/12/1932	16.00							
3	Amy	.00	04/23/1973	17.00							
4	Sierra	.00	09/19/1998	3.00							
5	Charlie	1.00	03/25/2001	1.00							
6											
7											

Data View / Variable View

SPSS Processor is ready

Figure 2.8 The Completed Data View Window

We will work with the data you compiled here again in the future. For now, however, save your data using these steps. Be sure to navigate to a folder that will not be erased, if the default is a temp file or if you are at a lab computer. Type in the *File Name* FAMILY and SPSS will add the extension .SAV to indicate that it is a saved data file:

File

Save As

File Name: FAMILY

Save

After you saved the file, SPSS will display the *Output Window*, with a written description saying the file has been saved. You can close that window and still have the *Data View* and *Variable View* remain operational. Congratulations, you have just defined variables, labeled variables, labeled values, designated variable types, entered information, and saved your data. These are the basic procedures in getting a data set operational.

Managing Data Sets: Dropping and Adding Variables, Merging Data Sets

Good data analysts become skilled at selecting data that are pertinent to their research questions. There are currently so much data available that the tasks of managing information can become overwhelming. Suppose that you are interested in looking at environmental issues and are compiling a new data set from existing data sources. As you compile data concerning recycling, car-pooling, etc., you come across homicide statistics. There might be a relationship between homicide and recycling, but you have to ask yourself at this point, does this new variable have anything to do with my main research question? If it does, keep it in your data set. If it does not, it may be advantageous to drop it from the data set entirely.

Dropping and Adding Variables

There are a number of ways to drop variables from the data. One approach is simply to go to the top of the *Data View* window, highlight the column of the variable that is not desired, and use the *Edit* command:

Edit
Cut

You can try this function by opening your data FAMILY.SAV and cutting the variable EDYRS. You will see this variable disappear from the *Data Editor* window. This variable can be restored by using:

Edit
Undo.

It is also possible to add a new variable to the data set using the command:

Data
Insert Variable

Of course, once you insert a new variable, you will need to define it and enter the appropriate values as described in this chapter. It is important to note that any changes in the data (such as adding or deleting variables) will be saved only if you save the data set before exiting the program. If you do not save the data, any changes will be lost. If you make substantial changes to a data set, it is prudent to save that file under a new name. As we work with data, we usually give the initial file a name such as "FAMILY.SAV." As we refine the data, adding or deleting variables, we save the new files with successive new names such as "FAMILY2.SAV," "FAMILY3.SAV," etc. One advantage of doing this is that it opens the possibility of backtracking and correcting any (and almost inevitable) mistakes made along the way. You might develop a different system, but it is important to keep a running record of changes to your data. It is also important to never copy over the original data set unless you are absolutely sure that you want all of the changes to remain. Once a data set is copied over, the information is changed forever.

Sometimes it is helpful to inspect a file, to see what types of variables are included, and to get a sense of their structures. To do this, perform the following command on your family data set:

File
Display Data File Information
Working File

Merging and Importing Files

All of the data provided with this text are in SPSS format, which makes your work much easier. However, at some point in your career, you may find it necessary to **import** statistical information from another format, such as Lotus, Excel, dBase, or ASCII. You will find SPSS has the capability of importing an Excel file, for example, with the command:

File
Open
Data
File name: Filename.xls
Files of Type: Excel (*.xls)
Open

In compiling the data for your analysis, we accessed different sources of data, selected variables that we considered pertinent for research projects of interest to students, and combined these sources through a process called **merging**.

Data
Merge Files
Add Variables

The *Add Variables* command combines two data sets by adding new variables to the end of a primary file. Suppose you wanted to add a file containing your family members' social security numbers to the data set you compiled earlier. You would need to first sort both files so that each family member occurs in the same order in each file using the command:

Data
Sort Cases
Sort by: PERSON

Once each file is ordered the same, SPSS can match cases so that the appropriate information is added to each case.

It is also possible to add new cases to the data set using the *Merge Files* command. Suppose, for example, that you wanted to combine the information about your family with information about other class members' families. As long as each file has the same variables, new cases (people) can be added using:

Data
Merge Files
Add Cases

There is no immediate need for you to import or merge files. However, in the future you may want to add some variables to the data set included with this text or construct an entirely new data set from existing data. If this is the case, you will probably need to use the merge or import procedures. If you have existing data in a format that SPSS cannot import, there are good software packages available, such as DBMS/Copy, that easily translate data files from one format to another.

Loading and Examining an Existing File

As you can tell, entering data can be labor intensive and time consuming. For this reason, many professional social science researchers allot this task to companies that specialize in data entry. Thankfully, there are also data already compiled by government agencies and by other researchers. In fact, there are a number of organizations (such as the Inter-University Consortium for Political and Social Research) that specialize in distributing data to their members. The U.S. Government makes considerable amounts of data available as well.

On the disk accompanying this text is a data set containing state level data, STATES07.SAV. These data are drawn from a number of sources such as the Department of the Census and the FBI. To examine this data set, use:

File
Open
Data
Look in:
File Name: STATES07.SAV
Open

Your computer has to search for these data in the correct place, so make sure that *Look in:* (located at the top of the window) directs your machine to search the appropriate drive.

The *Data Editor* window will fill with data from the 50 United States and the District of Columbia. Just as your family members comprised cases in the previous exercise, the states comprise cases in this data set. Your screen should look like Figure 2.9. You will notice that the data in the STATES07 data set have names that are a combination of initials and numbers. For example, the variable named DMS378 offers data on "Birth Rate per 1,000 Pop: 2005" (Place your cursor over a variable name and the variable label will pop up). The reason for this system is that we want you to be able to trace all of these data back to their original sources. In the case of the STATES07 data set, all of the data have been drawn from three books:

Crime State Rankings 2007 (Morgan Quitno 2007)
Health Care State Rankings 2007 (Morgan Quitno 2007)
State Rankings 2007 (Morgan Quitno 2007)

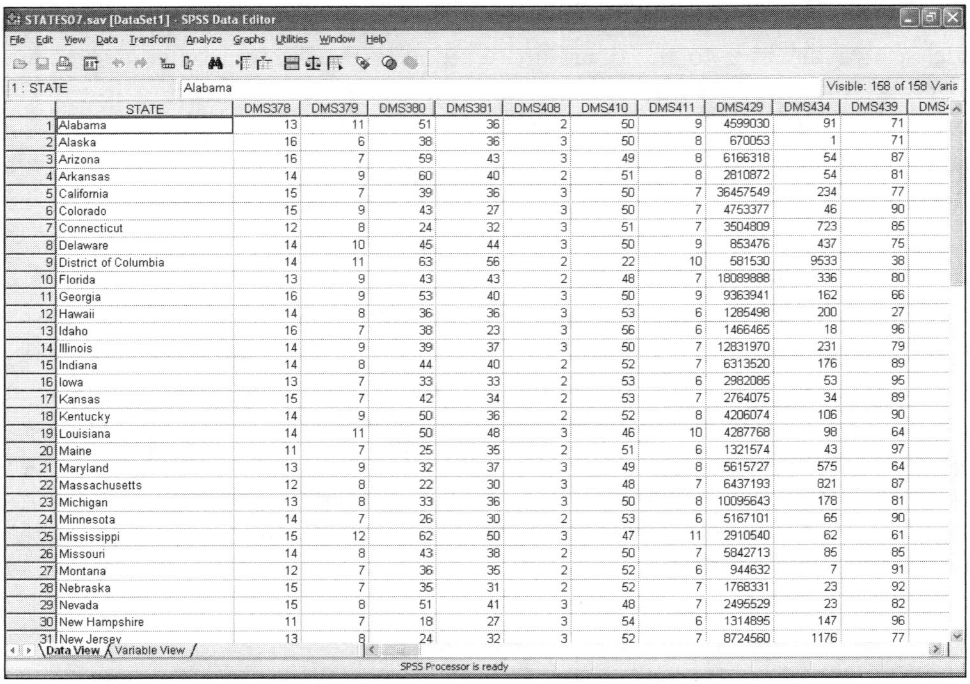

STATES07.sav [DataSet1] - SPSS Data Editor

File Edit View Data Transform Analyze Graphs Utilities Window Help

1 : STATE Alabama Visible: 158 of 158 Varia

	STATE	DMS378	DMS379	DMS380	DMS381	DMS408	DMS410	DMS411	DMS429	DMS434	DMS439	DMS4
1	Alabama	13	11	51	36	2	50	9	4599030	91	71	
2	Alaska	16	6	38	36	3	50	8	670053	1	71	
3	Arizona	16	7	59	43	3	49	8	6166318	54	87	
4	Arkansas	14	9	60	40	2	51	8	2810872	54	81	
5	California	15	7	39	36	3	50	7	36457549	234	77	
6	Colorado	15	9	43	27	3	50	7	4753377	46	90	
7	Connecticut	12	8	24	32	3	51	7	3504809	723	85	
8	Delaware	14	10	45	44	3	50	9	853476	437	75	
9	District of Columbia	14	11	63	56	2	22	10	581530	9533	38	
10	Florida	13	9	43	43	2	48	7	18089888	336	80	
11	Georgia	16	9	53	40	3	50	9	9363941	162	66	
12	Hawaii	14	8	36	36	3	53	6	1285498	200	27	
13	Idaho	16	7	38	23	3	56	6	1466465	18	96	
14	Illinois	14	9	39	37	3	50	7	12831970	231	79	
15	Indiana	14	8	44	40	2	52	7	6313520	176	89	
16	Iowa	13	7	33	33	2	53	6	2982085	53	95	
17	Kansas	15	7	42	34	2	53	7	2764075	34	89	
18	Kentucky	14	9	50	36	2	52	8	4206074	106	90	
19	Louisiana	14	11	50	48	3	46	10	4287768	98	64	
20	Maine	11	7	25	35	2	51	6	1321574	43	97	
21	Maryland	13	9	32	37	3	49	8	5615727	575	64	
22	Massachusetts	12	8	22	30	3	48	7	6437193	821	87	
23	Michigan	13	8	33	36	3	50	8	10095643	178	81	
24	Minnesota	14	7	26	30	2	53	6	5167101	65	90	
25	Mississippi	15	12	62	50	3	47	11	2910540	62	61	
26	Missouri	14	8	43	38	2	50	7	5842713	85	85	
27	Montana	12	7	36	35	2	52	6	944632	7	91	
28	Nebraska	15	7	35	31	2	52	7	1768331	23	92	
29	Nevada	15	8	51	41	3	48	7	2495529	23	82	
30	New Hampshire	11	7	18	27	3	54	6	1314895	147	96	
31	New Jersey	13	8	24	32	3	52	7	8724560	1176	77	

Data View / Variable View

SPSS Processor is ready

Figure 2.9 The Data Editor Display of STATES07.SAV

The authors of these books have compiled information from a wide variety of sources, and if you have any questions about the origins or limits of these indicators, their books will tell you where this information can be found. To simplify searches of the data for specific subjects, each variable has a two letter prefix, summarized here:

Variable Name Prefixes

CR - crime
DF – defense
DM - demography
EC - economy
ED - education
EM - employment
EN - environment
HT - health
LE - law enforcement
PO – politics
PV - poverty

The third letter in the variable name indicates the source of the data. C = *Crime State Rankings 2007*, H = *Health Care State Rankings 2007,* S = *State Rankings 2007.* The remaining numbers indicate the page of the text from which these data were drawn. For example, variable CRC222 (Reported Juvenile Arrest Rate per 100,000 for Drug Abuse Violations: 2005) can be interpreted as:

CR- Crime
C- *Crime State Rankings 2007*
222- Page 222.

On the surface, this approach to naming variables seems cumbersome. However, it offers a few advantages: it is systematic and allows the researcher to locate the variables quickly; it groups similar concepts together alphabetically; and it enables location of the original source of the data. After gaining experience in working with data, you will develop your own systems for naming and labeling variables. There really is no "right" or "wrong" variable name, but you will want the flexibility to add to the names. For example, if you need to modify a variable, you may want to add a suffix character to the end of the variable name (e.g., CRC222 might be converted into a new variable CRC222B).

To explore these data, you can move to the *Variable View Window* and observe the structures and labels for the different variables. You may also want to open the GSS07 data to examine the variables and their structures.

Summary

SPSS has the capacity to import data from other computer programs, to combine data from different sources into single data sets, and to facilitate the creation of entirely new data sets. In this chapter, you learned how to construct a data set and how to retrieve existing data. We also discussed data management and some of the skills to select appropriate data and develop systematic methods of saving and naming data sets. A well constructed data set is the foundation of all subsequent analysis, so particular attention needs to be paid to designating variable types, labeling variables clearly, and naming variables in a way so they are accessible to users.

Key Terms

Adding
Case
Categorical variable
Continuous variable
Count variable
Data view window
Dropping
Import
Menu bar
Merging
Missing value
Nominal variable
Ordinal variable
Scale variable
Output window
String variable
Syntax commands
Tool bar
Value label
Variable
Variable view window

Chapter 2 Exercises

Name______________________________________ Date____________

1. Create a new data set by polling 5 other people and asking the following questions:

RESPOND. What are the last four digits of your phone number?

EXER1. Do you exercise regularly? (Yes or No)

EXER2. How many hours a week would you say you exercise?

EXER3. Do you participate in team sports? (Yes or No)

EXER4. What is your favorite sport to play?

Create a new data set using these data and incorporating the following labels.

Variable Name	Type	Label	Value Labels	Measure
RESPOND	String20	Respondent Name	None	Nominal
EXER1	Numeric	Exercise Regularly?	0=no 1=yes	Nominal
EXER2	Numeric	Hours/Week of Exercise	None	Scale
EXER3	Numeric	Participate in Team Sports?	0=no 1=yes	Nominal
EXER4	String 20	Favorite Sport	None	Nominal

Enter the responses and save the data as "EXERCISE.SAV". Be sure to save it to a location that you have access to. Show that you have set up the file with the following command and follow your instructor's request, such as print the output:

File
Display Data File Information
Working File

2. Open the data STATES07.SAV. Examine the data set in the *Variable View Widow* and locate some variables that measure the following concepts. Recall that the variable names begin with a 2-letter code for the general topic of the variable. (Double click within the box of variables and scroll up or down to view all of the variables.)

A. Economic Prosperity

_______________ ____
Variable Name Variable Label

_______________ ____
Variable Name Variable Label

_______________ ____
Variable Name Variable Label

B. Substance Abuse

_______________ ____
Variable Name Variable Label

_______________ ____
Variable Name Variable Label

_______________ ____
Variable Name Variable Label

C. Educational Attainment

_______________ ____
Variable Name Variable Label

_______________ ____
Variable Name Variable Label

_______________ ____
Variable Name Variable Label

3. Your state's Department of Education hires you as a statistical consultant to help design a data set. They have already sent a survey to school superintendents, and need you to design their data set so they can begin entering information. Below is an example of a survey returned from one school district. Using the survey, answer the questions below, showing how you would design the data set.

1. Number of elementary schools:	8
2. Number of middle schools:	4
3. Number of high schools:	2
4. Average spending per pupil:	$7500
5. Student/Teacher ratio:	19/1
6. High School drop-out rate:	10%
7. Does the High School offer AP classes?	Yes
8.Type of district: (circle one)	Rural (Suburban) Urban

In this data set, what will be a case? Type of district

B. Design variable names, labels, and measurement types for each of the variables in the study.

	Variable Name	Variable Label	Value Labels	Measurement Type
1.	NUMELEM	# of Elementary schools		Scale
2.	NUMMID			Scale
3.	NUMHS			Scale
4.	AVGSPNDP			Scale
5.	STTCHRTO			Scale
6.	HSDROP			Scale
7.	HSAPOFR		1=Yes 0=no	Nominal
8.	DISTTYP		1=Rural 2=Sub 3=Urban	Nominal

4. Open the GSS04 data and examine the variable RACE. Can you identify the values and the value labels?

Value ________ Value label ____________________

Value ________ Value label ____________________

Value ________ Value label ____________________

Value ________ Value label ____________________

5. A novice researcher begins an analysis of the GSS04 data and wants to understand the distribution of income in the United States. She looks at the data in the data view window and sees values that range mostly from 1 to 13, but occasionally also sees the number 98 (you can do this by finding the variable INCOME98 in the GSS04 data). She explains to her instructor that the data must be wrong because almost no one has an income of "10." Her instructor tells her the data are fine, and asks her to come back when she can tell him how much a family income is for an observation that is listed as "10".

What does "10" in the INCOME98 variable indicate? 12500-14999

What is the reason for the novice researcher's confusion?

She doesn't understand that the labels are categories of income

6. Open the file FAMILY.SAV that you saved earlier and design 4 new variables to put into the data set and enter in the values. Try to vary the structure of the variables so that you include at least one scale variable, one ordinal variable, and one nominal variable. For example (don't use this one!), we could add the variable "FISH" with the label "Likes to Fish", the value labels "0=NO, 1-YES", and the measurement type "Nominal". Enter the responses and save the data as "FAMILY2.SAV". Show that you have set up the file with the following command and follow your instructor's directions, such as print the output:

    File
        Display Data File Information
            Working File

Chapter 3
Univariate Analysis: Descriptive Statistics

Overview

After data are entered and the variables created, it is possible to begin data analysis and make meaningful summary statements about variables. The first step in this process is understanding the characteristics of individual variables through univariate analysis. **Univariate analysis** explores the patterns of observations in individual variables in isolation from the other variables in a data set. This chapter demonstrates the generation and application of univariate statistics, including measures of central tendency and spread. We also show how to examine distributions of data with tables and graphs, including pie charts, bar charts, histograms, and box plots. While univariate analyses are often intended as building blocks for subsequent analyses, the findings generated by univariate analyses are oftentimes among the most interesting and important components of research projects.

Why Do Researchers Perform Univariate Analysis?

Some of the most interesting questions in social science research concern the frequencies of events. For individuals, some important questions are how often violence is experienced, how long families are dependent on welfare, or how often the typical worker changes jobs. And for societies, we might be interested in issues such as the volume of productivity, employment rates, incarceration rates, or other measures that enable the comparison of one place to another. While many studies seek to analyze the relationships between variables, some of the most important questions precede those analyses by simply asking "how much (or how often) does this type of observation occur?"

Most social science research is about relationships between variables, such as those between education and crime, income and education, or religion and suicide. If this is so, why study variables in isolation from one another? The foremost reason is that univariate analysis is very informative. For example, to fully understand whether education affects crime requires knowing how educated people are—the percentages of the population who drop out of high school, graduate from high school, and go on to graduate from college. If crime is a concern, what are the overall crime rates and how much do they vary in society? Knowing the characteristics of individual variables is essential for understanding the implications of any relationships between them.

A second reason for univariate analyses is that they inform researchers of the types of advanced statistical procedures that can be used. As we show later in this book, these decisions depend on the characteristics of individual variables, which are revealed through univariate analysis. A final reason is that univariate analysis reveals limitations in the data, including patterns of missing data and any data entry errors.

In sum, in this chapter our concern is considering how to understand the characteristics of all of the variables that you may eventually be interested in linking with each other. If you learn to do univariate analysis well, you will be able to determine if any individual case is "typical" or "atypical," and how values of observations are arrayed on a continuum.

Exploring Distributions of Scale Variables

Throughout this chapter, we will be using the STATES07 and the GSS04 data to demonstrate univariate analysis. The variables selected in the STATES07 data illustrate how SPSS can create statistics and graphs for scale variables, and the GSS04 data do the same for categorical variables. At this point, start SPSS and load the STATES07 data. Recall that you can do this through the menus following the commands:

File
  Open
    Data
      File Name: STATES07.SAV
      Open

Our first concern is to examine distributions of scale variables. A **distribution** is an account of all of a variable's values. The most rudimentary way of describing a variable's distribution is to make a list of all the values for all the cases. Although this method is comprehensive, it can be overwhelming if there are many cases, and patterns are often obscured. As you will see, SPSS offers a number of functions to better summarize distributions.

Listing, Summarizing, and Sorting Observations

Try this exercise. Using the STATES07 data, summarize the variables CRS38 (Rape Rate per 100,000: 2005) and STATE (the 50 U.S. states and Washington D.C.). In this analysis, the variable STATE simply labels the observations, easing our interpretation of univariate output for CRS38, the rape rates of individual states.

Analyze
  Reports
    Case Summaries
      Variables: STATE
        CRS38

Case Summaries[a]

	STATE State	CRS38 Rape Rate per 100,000: 2005
1	Alabama	34.3
2	Alaska	81.1
3	Arizona	33.8
4	Arkansas	42.9
5	California	26.0
6	Colorado	43.4
7	Connecticut	20.0
8	Delaware	44.7
9	District of Columbia	30.2
10	Florida	37.1
11	Georgia	23.6
12	Hawaii	26.9
13	Idaho	40.4
14	Illinois	33.7
15	Indiana	29.6
16	Iowa	27.9
17	Kansas	38.4
18	Kentucky	34.0
19	Louisiana	31.4
20	Maine	24.7
21	Maryland	22.6
22	Massachusetts	27.1
23	Michigan	51.3
24	Minnesota	44.0
25	Mississippi	39.3
26	Missouri	28.0
27	Montana	32.2
28	Nebraska	32.9
29	Nevada	42.1
30	New Hampshire	30.9
31	New Jersey	13.9
32	New Mexico	54.1
33	New York	18.9
34	North Carolina	26.5
35	North Dakota	24.2
36	Ohio	39.8
37	Oklahoma	41.7
38	Oregon	34.8
39	Pennsylvania	28.9
40	Rhode Island	29.8
41	South Carolina	42.5
42	South Dakota	46.7
43	Tennessee	36.4
44	Texas	37.2
45	Utah	37.3
46	Vermont	23.3
47	Virginia	22.7
48	Washington	44.7
49	West Virginia	17.7
50	Wisconsin	20.6
51	Wyoming	24.0
Total N	51	51

a. Limited to first 100 cases.

Figure 3.1 A Case Summary of Two Variables

Case Summaries[a]

	STATE State	CRS38 Rape Rate per 100,000: 2005
1	New Jersey	13.9
2	West Virginia	17.7
3	New York	18.9
4	Connecticut	20.0
5	Wisconsin	20.6
6	Maryland	22.6
7	Virginia	22.7
8	Vermont	23.3
9	Georgia	23.6
10	Wyoming	24.0
11	North Dakota	24.2
12	Maine	24.7
13	California	26.0
14	North Carolina	26.5
15	Hawaii	26.9
16	Massachusetts	27.1
17	Iowa	27.9
18	Missouri	28.0
19	Pennsylvania	28.9
20	Indiana	29.6
21	Rhode Island	29.8
22	District of Columbia	30.2
23	New Hampshire	30.9
24	Louisiana	31.4
25	Montana	32.2
26	Nebraska	32.9
27	Illinois	33.7
28	Arizona	33.8
29	Kentucky	34.0
30	Alabama	34.3
31	Oregon	34.8
32	Tennessee	36.4
33	Florida	37.1
34	Texas	37.2
35	Utah	37.3
36	Kansas	38.4
37	Mississippi	39.3
38	Ohio	39.8
39	Idaho	40.4
40	Oklahoma	41.7
41	Nevada	42.1
42	South Carolina	42.5
43	Arkansas	42.9
44	Colorado	43.4
45	Minnesota	44.0
46	Delaware	44.7
47	Washington	44.7
48	South Dakota	46.7
49	Michigan	51.3
50	New Mexico	54.1
51	Alaska	81.1
Total N	51	51

a. Limited to first 100 cases.

Figure 3.2 A Sorted Case Summary

Your output should look very much like Table 3.1, which shows the states listed in alphabetic order (the original file format) and their Rape Rates (per 100,000 people in the state).[6] This output is useful in that we can quickly find the rape rates for individual states. However, as the data are currently structured, it takes some effort to find which states have the highest and lowest rates, or to observe the full range of values (assess the difference between the highest and lowest rates). Also, we do not have a sense of the most typical rate, and/or the points at which rates become exceptionally high or low. We need a less cumbersome approach if we are going to understand the general patterns of rape in the United States.

One way to refine the analysis and gain further appreciation for the distribution of rate rapes within the United States is to **sort** the data. **Sorting** – arranging the data in ascending or descending orders - can be a very handy feature for performing other operations. For instance, it can be used during **data cleaning**, the phase of a project when researchers examine data for potential errors and make corrections. Sorting is also needed to split files into groups for separate analyses, or to prepare data for merging with other data sets. The STATES07 data are already sorted alphabetically by the variable STATE, but let us now impose a new order, by sorting by the variable CRS38:

Data

Sort Cases

Sort By: CRS38

OK

If you repeat the *Case Summaries* for variable CRS38 and STATE, you should observe the states in ranked order of the incidence of rape and output should look like Figure 3.2. The advantage of sorting the data is that you can now quickly identify which states have the highest and lowest rates. Here we see a surprising finding – New Jersey, West Virginia and New York have remarkably low rate rapes compared to Alaska, New Mexico and Michigan (which have rates that are 4 to 5 times higher). What is it about these places that produces, (or discourages) the incidence (or potentially the reporting) of rape? We leave that issue to you to explore in subsequent chapters and potentially in your own research project (see Chapter 10).

Although we now know more about the variation in rape rates than before, sorting does not yield meaningful summaries. When data sets have many observations, graphs or summary statistics are needed to represent the data so as not to rely on comparisons of individual cases.

At this point, sort the data back so that it is ordered by the variable STATE. When you get to the "sort cases" window this time, either click "reset" or move CRS38 back to the main menus since SPSS saves the most recent setting during any one session. Then resort the cases back to their original order with the variable STATE.

---

[6] Note: If your output does not look like Figure 3.1, reconfigure the output labels to display both names and labels, as well as values and labels. This can be done by going to

Edit

Options

Output Labels

Data
Sort Cases
Sort By: STATE
OK

Note that the data can remain in any sorted order, and it will not affect your analyses of the data set, only the way the cases in the *Data Editor* window are arrayed.

Histograms

A **histogram** is a graphic summary of the distribution of a scale variable. In a histogram, classes of values are listed along the horizontal axis. The height of the bar above each class indicates the number of cases with values in that class which makes more apparent any distributional patterns in the data. Using the STATES07 data, and the interactive graph builder, we can produce a histogram representing the distribution of the rape rates.

Graphs
Interactive
Histogram
Normal curve
Assign Variables
$count:
CRS38
OK

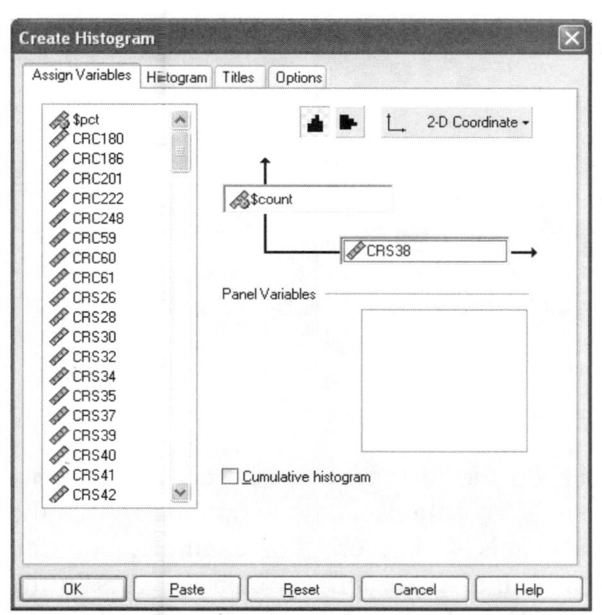

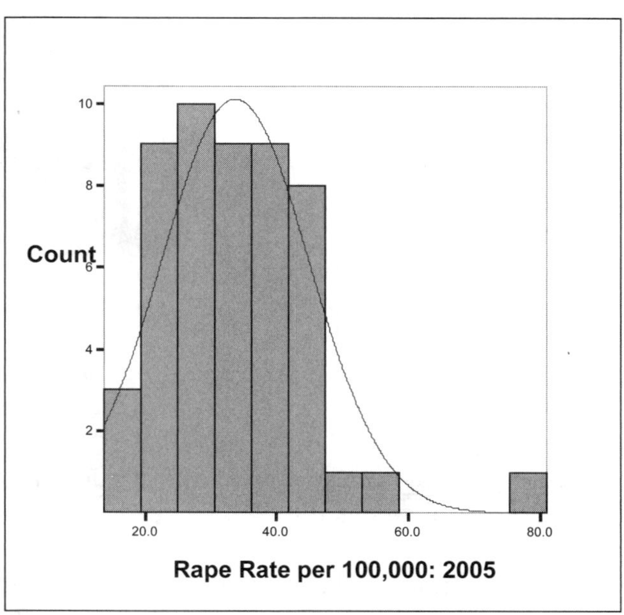

Figure 3.3 Using the Interactive Chart Builder to Create a Histogram

Notice the information the histogram gives the analyst. First, we observe that most states cluster in the rate range of 20-50 (per 100,000). The overlapping curve, imputed from the SPSS program, indicates that the distribution is somewhat bell shaped (as we discuss later, this general pattern is needed for some types of analyses). But we also observe that there is one state (Alaska—refer back to the sorted list) that stands far apart from the rest of the nation. This opens an analytic puzzle, should we keep Alaska in any subsequent analysis of the contextual factors associated with increased incidences of rape, or should we treat it as a separate and unique case? Perhaps, if you pursue this analysis further, you could try analyzing the data both ways and observe the outcomes.

One convenient feature of SPSS is that it automatically groups the values into classes for histograms. It is possible, however, to change the width of the classes into more refined groupings by stipulating the number of bars to be represented in the histogram. In the case of the histogram in Figure 3.3, we allowed SPSS to automatically generate what it believes to be an appropriate number of bars. We could, however, override this function and command it to increase or decrease the number of bars by invoking the **Interactive Chart Editor** commands available in SPSS. You will inevitably want to explore these options, especially when it comes time to publish some of the graphs you produce.

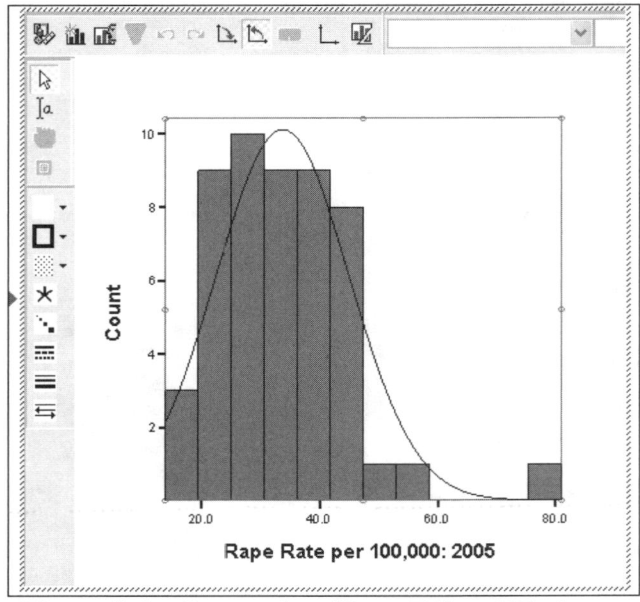

Figure 3.4 Interactive Chart Editor Window

To access the *Interactive Chart Editor*, double click on the histogram in the *Output* window (Figure 3.4). This opens the SPSS *Interactive* window. You will observe new options next to the graph, which allow you to change titles, axis labels, axis scales, etc. For example, you can specify the number of bars represented on the graph with a single left click on the bars in the *Interactive* window (all of the bars will become highlighted). Then right click and select *Interval Tool*, which will enable you to increase or decrease the range of values represented by each bar. After you complete making any edits to the graph, click outside of the editor box and the changes will be made to your output.

Shapes of Distributions

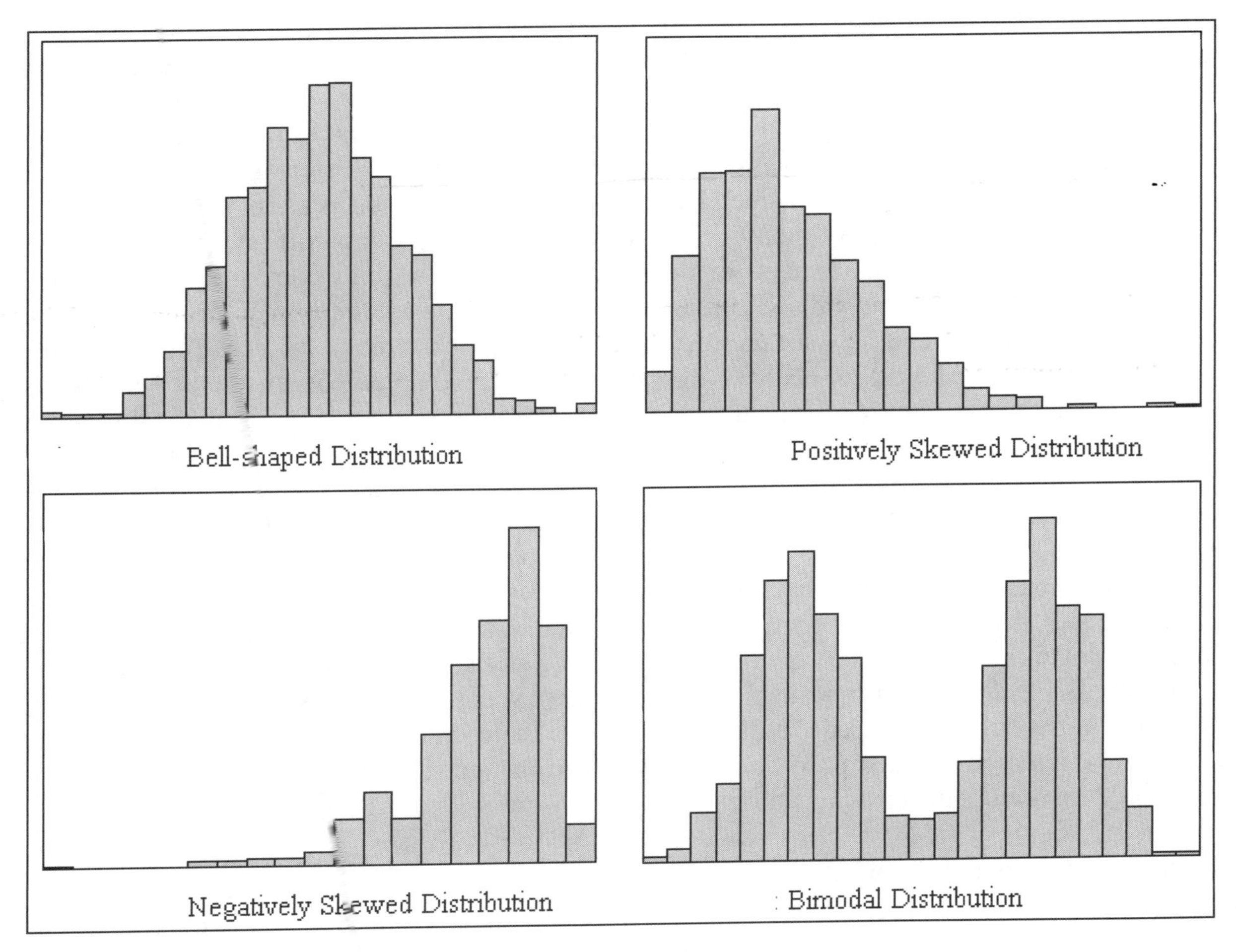

Figure 3.6 Histograms of Normal, Skewed, and Bimodal Distributions

Histograms represent distributions and they are often summarized with written descriptions of their shapes. There are four common shapes, illustrated in Figure 3.6. A **bell-shaped distribution** is symmetric—both sides look pretty much the same. It has the highest frequency of cases in the middle, and the frequencies taper off at the high and low ends of the distribution, creating "tails" with few cases.

A **positively skewed distribution** looks like a bell-shaped distribution that has been shoved to the left, with a tail that points in the positive direction on the horizontal axis. There are many cases with low values and there are few cases with high values. A common example of a variable with a positively skewed distribution is the salaries at a company. There are often many workers with relatively low salaries, and fewer and fewer people making higher and higher salaries.

A **negatively skewed distribution** has the reverse pattern. These distributions have many cases with high values and few cases with low values. If people rated their health on a 10 point scale, with 10 indicating very good health and 0 indicating poor health, the distribution would be negatively skewed, as most people are in good health.

A **bimodal distribution** is also symmetric, but the highest frequencies are at both ends, with few cases in the middle. Daytime traffic accidents fit this pattern, with most accidents occurring in the early mornings and late afternoons during the commute to and from work. Although these four patterns are common, there are many other distributions with other patterns or no discernible pattern at all.

When describing the shape of a distribution, it is also important to observe if any outliers are present. An **outlier** is an observation so far removed from the cluster of other observations that it is considered an extreme value. How extreme does an observation have to be to be considered an outlier? There are many different ways of defining a cutoff and researchers will need to consider the particular aspects of their data, including sample sizes and the distance of the outlier from the rest of the values. But one needs to always pay attention to observations that are positioned quite differently than the rest of the values, which make the case (or small set of cases) extreme in their leverage. As we will discuss in future chapters, because outliers can have a strong impact on the values of statistics, they can undermine conventional statistical methods. Some researchers will automatically discard any outliers from a data set as "unrepresentative." But this may not be an altogether satisfactory strategy because outliers can be among the most substantively important observations. Knowing that outliers exist is the first step in determining the best way to handle them.

Sometimes a researcher can determine that an outlier is a result of an error, either from a measurement or data entry. For example, one of the authors was once asked to analyze a data set of women's weights, and there was one woman who was listed as weighing 19 pounds. As it is nearly impossible for an adult woman to weigh only 19 pounds, we easily concluded that this was a data entry error and traced the data back to find the correct value. It is always a good idea to examine outliers in an attempt to find data entry errors, and if these errors cannot be remedied, they should be removed from the data set.

Measures of Central Tendency

Summarizing data through graphs helps researchers understand the nature of the variable distribution. However, some situations require a concise summary of a variable's central values. One of the most important pieces of information concerns the center point in the variables' distribution. Three **measures of central tendency** are common to data analysis: the mode, mean, and median. All share in common the objective of locating a midpoint in the range of observed values. For example, people come in all sizes, ranging from the very small to the very large, but sometimes it is helpful to know the average or typical height of a person as a benchmark from which to judge the heights of individuals. As we discuss below, the "average," is not always the best indicator of what is "typical," nor is the most commonly observed value necessarily the midpoint in the data. The interesting challenge is determining which measure of central tendency best describes the midpoint, and deciding what measure is the best one to report.

First, let us consider the **mode**, a statistic that reports the most frequently observed value in a variable. Of the three measures of central tendency, the mode is probably used least in formal reports, but is used most in understanding distributions in data sets. Consider describing the criminal behavior of the typical teenager and we find that the "average" teen reports having committed 3 serious crimes. But we could think of the "typical" teen in a different way, by asking "what is the most common number of crimes reported by teens?". It is entirely possible that most teens commit no serious crimes, and that the average crime rate is elevated by a minority of hard core individuals who commit many crimes, thus elevating the group average. Or to take another example, the typical American family has 2.3 children. Of course one cannot have two and one thirds children, which would make any family accomplishing this seemingly impossible feat atypical! In both of these cases, it may make more sense to say that the "typical" observation is very different than the mathematical average.

SPSS calculates the mode using the frequencies command. Using the STATES07 data, run a frequency on the number of prisoners executed from 1977-2005 (CRC77):

Analyze
Descriptive Statistics
Frequencies
Variable: CRC77
OK

Your findings should look like Figure 3.6. If we were to describe the "typical" number of executions in a state during this period, the mode of 0 could be a good measure to report because 18 states (35.3%) executed no prisoners. The next most common number of executions was 1, performed by 6 states (11.8%). These statistics show that, in most states, executions are uncommon. However, one state executed 355 prisoners. Perhaps you can guess which state this is. See if you were correct using *Sort* then *Case Summaries*.

CRC77 Prisoners Executed: 1977 to 2005

		Frequency	Percent	Valid Percent	Cumulative Percent
Valid	.0	18	35.3	35.3	35.3
	1.0	6	11.8	11.8	47.1
	2.0	3	5.9	5.9	52.9
	3.0	2	3.9	3.9	56.9
	4.0	1	2.0	2.0	58.8
	5.0	1	2.0	2.0	60.8
	6.0	1	2.0	2.0	62.7
	7.0	1	2.0	2.0	64.7
	11.0	1	2.0	2.0	66.7
	12.0	2	3.9	3.9	70.6
	14.0	1	2.0	2.0	72.5
	16.0	1	2.0	2.0	74.5
	19.0	1	2.0	2.0	76.5
	22.0	1	2.0	2.0	78.4
	27.0	2	3.9	3.9	82.4
	34.0	1	2.0	2.0	84.3
	35.0	1	2.0	2.0	86.3
	39.0	2	3.9	3.9	90.2
	60.0	1	2.0	2.0	92.2
	66.0	1	2.0	2.0	94.1
	79.0	1	2.0	2.0	96.1
	94.0	1	2.0	2.0	98.0
	355.0	1	2.0	2.0	100.0
	Total	51	100.0	100.0	

Figure 3.6 Frequencies Table of CRC77

The **mean** is generally what people refer to when they say "average." It is the arithmetic average of all the observations of a variable. In all likelihood, your grade point average (GPA) is calculated using a variation of the mean, in which your college averages together all of your course grades (all of which may be weighted differently depending on how many credits hours each course is). This ultimately produces an average that indicates the central tendency of all of your work.

$$\text{MEAN} = \frac{\text{SUM OF THE VALUES OF ALL OBSERVATIONS}}{\text{TOTAL NUMBER OF OBSERVATIONS}}$$

The mean is probably the statistic most reported by social scientists. It is especially useful when discussing research findings with the general public, who are less likely to be conversant in advanced statistics. For example, newspaper reports of average SAT scores usually refer to the mean SAT scores of all students. Many of the indicators in the STATES07 data sets are based upon means, such as the average amount spent per pupil in each state. Another reason why the mean is important is that it is the basis of many of the statistical procedures used to examine relationships between variables.

You have already found one way of locating the mean, by using the *Descriptives* command. Locate the mean of variable CRC77 (Prisoners Executed 1977-2005) using the *Descriptives* command. You should find the mean number of executions to be 19.627.

Analyze
Descriptive Statistics
Descriptives
Variable: CRC77
OK

Finally, the **median** is the value that separates the highest 50% of the cases from the lowest 50% of the cases. It is the center value in a rank ordered data set. The median is an especially useful measure of central tendency to describe the center point in highly skewed data. For example, there is a small minority of individuals, like Bill Gates, who have incomes that are so far removed from the rest of society that they pull the mean (mathematical average) income up. For this reason, the median, or the half-way point in the distribution, is oftentimes used to mark income trends or differences between groups in society. The SPSS *Explore* command finds both the median and the mean of a variable. Take a few minutes to try this. Generate *Explore* output for variable CRC77 (place CRC77 in the *Dependent Variable* box in the *Explore* window). Your output should include a table similar to Figure 3.7.

Analyze
Descriptive Statistics
Explore
Dependent List: CRC77
OK

Descriptives

			Statistic	Std. Error
CRC77 Prisoners Executed: 1977 to 2005	Mean		19.627	7.3607
	95% Confidence Interval for Mean	Lower Bound	4.843	
		Upper Bound	34.412	
	5% Trimmed Mean		11.080	
	Median		2.000	
	Variance		2763.198	
	Std. Deviation		52.5661	
	Minimum		.0	
	Maximum		355.0	
	Range		355.0	
	Interquartile Range		19.0	
	Skewness		5.491	.333
	Kurtosis		34.337	.656

Figure 3.7 Descriptives Output of CRC77 Generated by *Explore*

In the *Descriptives* table we find the median value for CRC77 is 2. The median center point in the data, as indicated by the median, had two executions from 1977-2005. Now reflect back on the purpose of reporting a central tendency. Which measure would be better for reporting the typical frequency of executions? Would it be better to report that the "average" state executed nearly 20 prisoners over the course of the past 28 years (the mean), or that the most common practice was to execute no prisoners (the mode), or very few (the lower tail of the distribution and the median)? There is not necessarily a correct answer to this question, as each indicator approaches the issue of identifying what is "typical" differently. Each measure tells something different about the rate of executions.

Measures of Spread

As the discussion above suggests, understanding data involves not only locating center points, but also the ranges of values present. For example, although the median individual income in the United States in 2004 was $29,931 very few people actually earned that exact amount. A interesting analysis is how closely earnings cluster around that figure, and how far they depart. As an example, consider that in 2005 the CEO of Wal-mart made $15,681,507, which is *524 times* the median income of all workers in the United States. The richest fifth of families earned half (48%) of the aggregate income earned by all families in the United States, while the poorest fifth of American families earned only 4% of the collective earnings (Sweet and Meiksins 2008). All of these observations about income disparities require considering not simply the center point, but also the range of values. To gain a sense of the spread of data, there are a variety of measures.

For example, the *Descriptives* command reports the **standard deviation,** which measures the clustering of data around the mean. A large spread of the data away from the mean results in a large standard deviation. A tight concentration of data around the mean results in a small standard deviation. If a variable has a bell-shaped distribution, 68% of the cases will have values that fall between the values that are one standard deviation below the mean and one standard deviation above the mean. Extending the range to two standard deviations above and below the mean captures 95% of the cases and extending it to three standard deviations from the mean captures about 99% of the cases. Therefore, a researcher who knows the mean and the standard deviation has enough information to gauge if any case is typical or atypical relative to the vast majority of cases.

Descriptives

			Statistic	Std. Error
EDS128 Estimated Public High School Graduation Rate: 2006	Mean		72.014	1.3297
	95% Confidence Interval for Mean	Lower Bound	69.343	
		Upper Bound	74.684	
	5% Trimmed Mean		72.261	
	Median		73.400	
	Variance		90.167	
	Std. Deviation		9.4956	
	Minimum		46.0	
	Maximum		90.1	
	Range		44.1	
	Interquartile Range		13.7	
	Skewness		-.390	.333
	Kurtosis		.048	.656

Figure 3.8 Descriptives (Generated by *Explore*) of EDS128

As a point of illustration, analyze variable EDS128 (Estimated Public High School Graduation Rate: 2006). Using the *Explore* command, you will find that this variable has a mean of 72.01 and a standard deviation of 9.49 (Figure 3.8). If plus or minus one standard deviation from the mean captures 68% of the cases, we know that 68% of the cases of variable EDS128 fall within the range 62.52 to 81.5. How did we arrive at this?

Mean – St Dev = Low Value	Mean + St Dev = High Value
72.01 – 9.49 = 62.52	72.01 + 9.49 = 81.5

At this point, pull back from the calculation of these statistics for a minute and reflect on their meaning. We began this chapter by suggesting that some of the most interesting aspects of any study come from simply describing the distribution of individual variables. Before looking at these data, did you realize that there was such wide variation among states? If you use the *Case Summary* command described earlier in this chapter, you will observe that the "best" state is Minnesota, which has a graduation rate of 90.1%, but the worst is the District of Columbia, with a rate of 46.0%.[7] Suppose we do not know anything about graduation rates except for the standard deviation and the mean, and need to consider West Virginia in a comparative context.

[7] Variable EDS128 was calculated by the U.S. Department of Education by comparing the estimated number of public high school graduates in 2005-2006 with 9th grade enrollment in 2002-2003. Because the data exclude ungraded pupils and have not been adjusted for interstate migration or switching to or from private schools and students graduating late, it likely over-estimates the drop-out rate.

Would its graduation rate of 72.9 indicate that its school system is performing relatively well or poorly?

Because the standard deviation is based on the mean, skewed distributions and outliers greatly affect it. Outliers, for example, introduce large distances between themselves and the mean, inflating the standard deviation. It can make the data as a whole seem more dispersed than it is for the vast majority of cases. Because of this, the standard deviation is most useful for describing symmetric distributions. If you produce a histogram of EDS128 (you can try this), you will see that it satisfies this assumption.

Two other common measures of spread are the **range** and the **interquartile range** (IQR). The range is the distance spanned by the data set. The range for the rate of graduation rates is 90.1- 46, which equals 44.1 (this number is reported in Figure 3.8). The IQR is the distance spanned by the middle 50% of the cases. Although it is slightly less intuitive than the range, its advantage is that outliers do not affect it, since the ends of the distribution are not used in its calculation. The IQR is calculated from parts of the **five number summary**, which is a set of five cutoff points in the distribution. These cutoff points include the minimum, the 25th percentile, the median, the 75th percentile, and the maximum. These values split the data set into four equal pieces. The lowest 25% of the cases will have values between the **minimum** and the **25th percentile**. The next 25% of cases will have values between the 25th percentile and the median. The next 25% of cases will have values between the median and the **75th percentile**, and the highest 25% of cases will have values between the 75th percentile and the **maximum**. The IQR is the value of the 25th percentile subtracted from the value of the 75th percentile. Since 25% of cases are below the 25th percentile and 25% of cases are above the 75th percentile, the IQR covers the middle 50% of the data set.

As reported in the descriptives in Figure 3.8, the interquartile range for the graduation rates is 13.7. If we take the median (73.4) and add 6.85 (half of the interquartile range) and subtract 6.85, we capture the middle 50 percent of cases, which will fall within the range of 66.55 and 80.25. This range can be illustrated graphically with box plots.

Box Plots

Okay, your head may now be clouded by all of the discussion of percentiles and ranges. Lets try to clear it up with a visual depiction of data based on the five number summary. Figure 3.9 shows an illustration of a **box plot**, identifying how these innovative graphs show medians, percentiles, interquartile ranges, extended ranges, and outliers. The main box (the shaded portion) extends from the 25th to the 75th percentiles. Recall, the distance between these two points is called the interquartile range (IQR). The line inside the main box represents the median. Remember that 50% of the cases fall above and 50% of the cases fall below this line. The "whiskers" (the thin lines extending out from the main box) extend 1.5 times the IQR from the top and bottom of the box. Those cases that are outside the whiskers are termed outliers and are signified by small circles.

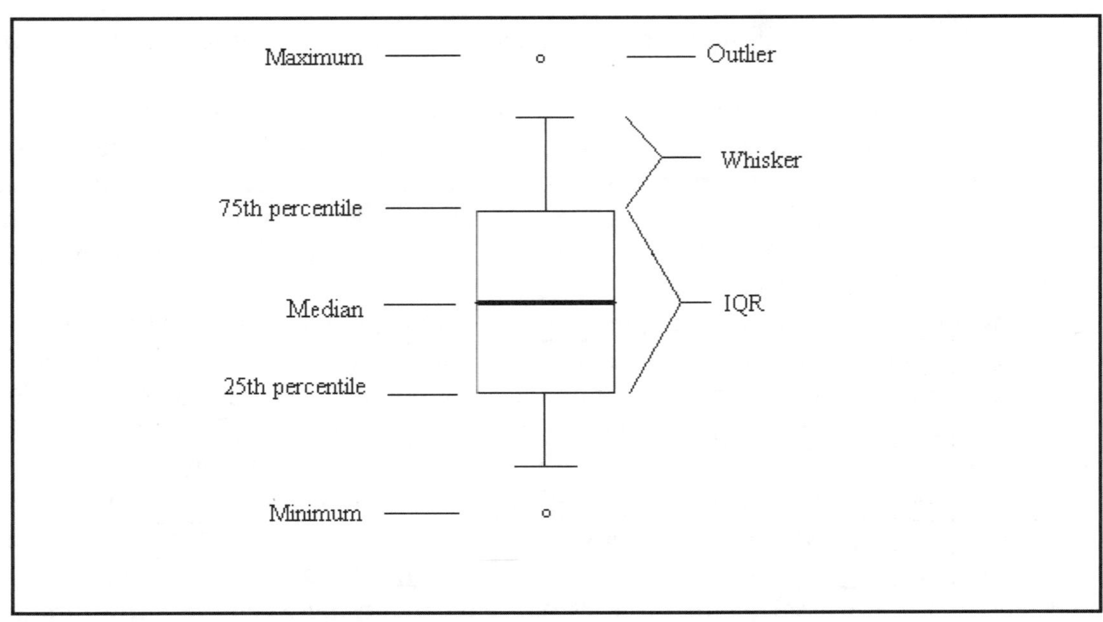

Figure 3.9 Box Plot Diagram

With this information in hand, we can create and interpret a box plot representing the distribution of high school graduation rates (EDS128). You can generate this box plot using the *Explore* command. If a box plot is not displayed when you run *Explore*, examine whether "*Display:* Both" is selected in the *Explore* window. A box plot of graduation rates is shown in Figure 3.10.

Analyze
Descriptive Statistics
Explore
Display: Both
Dependent List: EDS128
OK

Note in Figure 3.10 that the shaded box begins at approximately 66 and ends at approximately 80. From this box plot, we know that 50% of the cases fall between these two values and that this corresponds with the already identified interquartile range of 13.7. Since the median falls fairly close to (but not exactly in) the center of the box, and the wiskers are of modestly different sizes, we know the distribution is slightly skewed, but also on the whole quite symmetric. We do not observe any outliers, indicating that no states stand alone in the distribution outside the general range of the other states. Candidly, we like box plots, use them frequently, and wonder why they are not used more often!

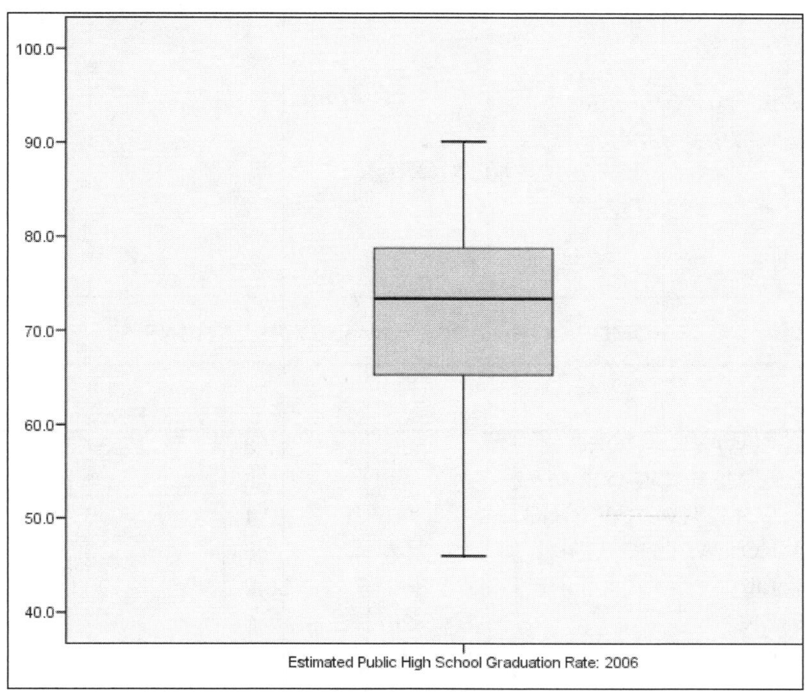

Figure 3.10 Box Plot of EDS128

Exploring Distributions of Categorical Variables

Recall that categorical variables are structured to represent classifications of observations. For individuals, these classifications might be on the basis of gender, race, or other socially constructed divisions that distinguish one person from another, and groups them with others who share similar qualities. Distributions of categorical variables can be summarized through a **frequency distribution table**. Like a histogram, a frequency distribution table gives the values of a variable and the number of times each value occurs among the cases in the data set. An advantage of this type of output is that it preserves information about all the values, but reduces output to a succinct list. To demonstrate how SPSS produces a frequency distribution for categorical data, open the GSS04 data. These data, recall, represent interviews with people, and measure attitudes and behaviors of individuals, not the characteristics of places.

File

Open

Data

File Name: GSS04.SAV

Open

In this exercise, we examine the degree to which respondents agreed with the statement "Do you think sexual relations between two adults of the same sex is always wrong, almost always wrong, wrong only sometimes, or not wrong at all?" A frequency distribution provides a useful summary of the responses. Try this:

Analyze
Descriptive Statistics
Frequencies
Variable: HOMOSEX
OK

HOMOSEX HOMOSEXUAL SEX RELATIONS

		Frequency	Percent	Valid Percent	Cumulative Percent
Valid	1 ALWAYS WRONG	500	17.8	57.6	57.6
	2 ALMST ALWAYS WRG	41	1.5	4.7	62.3
	3 SOMETIMES WRONG	60	2.1	6.9	69.2
	4 NOT WRONG AT ALL	267	9.5	30.8	100.0
	Total	868	30.9	100.0	
Missing	8 DK	24	.9		
	9 NA	1920	68.3		
	Total	1944	69.1		
Total		2812	100.0		

Figure 3.11 Frequency Distribution of a Categorical Variable HOMOSEX

Your output should look like Figure 3.11. This table gives a lot of information about the distribution. The first column of the table lists all the possible values of the variable. In this example, they are "Always Wrong", "Almost Always Wrong", "Sometimes Wrong", and "Not Wrong at All." Of the 2,812 people who were interviewed, 1,920 were not asked this question (NA) and 24 responded that they did not know (DK). Those cases are coded as missing. The second column, *Frequency,* indicates how many respondents chose each answer. Notice that the frequencies of people who chose these options add up to 868 (500+41+60+267 = 868), but that only about one in three people in the study offered a usable response.

The next three columns show three different types of percentages. The *Percent* column gives the percentage for each response out of all 2,812 interviewees. This gives an indication of how much of the data are usable. In contrast, the *Valid Percent* column gives the same percentages out of the 868 people who responded to the question. This would be the percentage offered in a report concerning attitudes toward homosexual sex. Finally, the last column, *Cumulative Percent,* displays the collective number of people who responded to the question, summing row by row in the table. Among those who offered answers to the homosexuality question, clearly most expressed some concern that it can be "wrong." But perhaps most important is the observation of how many report that it is "always wrong" and how many report that it is "not wrong at all". This appears to be a very polarizing question, with strong views being common on both ends of a continuum of attitudes.

Frequency tables can also be used to analyze scale data, such as rape rates, but only if the number of values is limited. The execution example in Figure 3.6 illustrates an appropriate frequency distribution of a scale variable. Because 0 is such a common value, there are only 23 unique values listed. In contrast, the rape rate example in Figure 3.2 illustrates a less useful approach to applying a frequency distribution. Because of the decimals, almost every value is unique. A frequency distribution that lists almost every value as occurring once does not summarize data efficiently. Instead, it is more useful to group the numerical values into classes and count the frequency of each class. A **class** is a grouping of values that are reasonably comparable, and, as such, classes become very similar to ranked categories of scale data. Although some information is lost when a class is created, the data are simplified so that researchers can interpret them. SPSS does not automatically create classes for a frequency distribution table. In order to do this, you have to recode the variable into a new variable. A discussion of recoding is in Chapter 4. SPSS will, however, automatically construct classes for histograms

Pie Charts

The **pie chart** is one of the most popular methods of graphing distributions and is still used heavily in marketing and business presentations of data. One reason for its popularity relates to the underlying conceptualization of the pie chart, that a limited resource (the pie) is being distributed (the wedges) to a definable population. A pie chart quickly reveals if any resource (or attitude, or behavior) is distributed proportionately, depending on the size of the wedges.

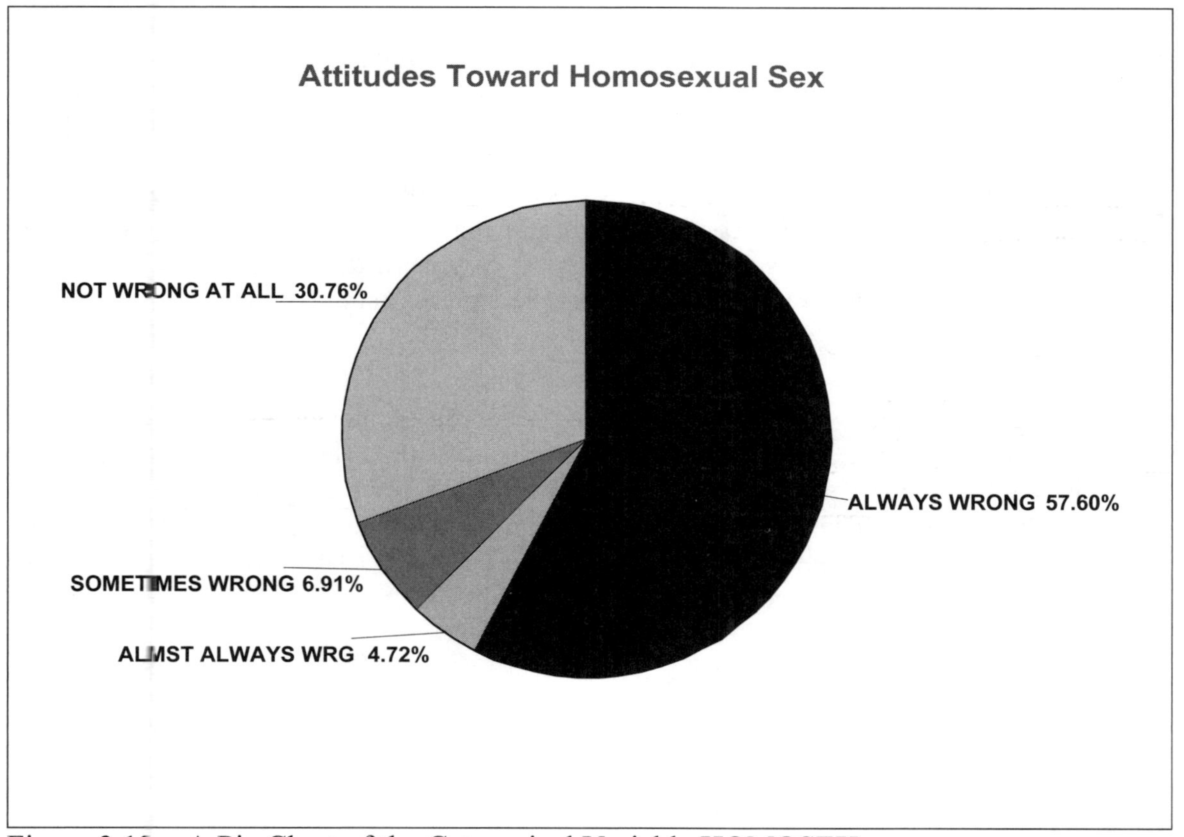

Figure 3.12 A Pie Chart of the Categorical Variable HOMOSEX

To make a pie chart representing attitudes toward homosexuality, illustrated in Figure 3.12, perform the following:

Graphs
Interactive
Pie
Simple
Assign Variables
Slice By: HOMOSEX
Slice Summary: $count
Titles
Chart Title: Attitudes Toward Homosexual Sex
Pies
Slice Labels: Category, Percent
All Outside
OK

(If missing values are incorporated into the pie chart, check *Options.*)

You will observe that the pie chart produced on your screen looks slightly different than that in your text. In order to format the pie chart to suit black and white print, we took advantage of some of the *Chart Editor* commands available in SPSS. You will inevitably want to explore these options, especially when it comes time to publish graphic presentations.

Although pie charts remain popular in business and marketing, they have largely fallen out of favor with social scientists. One concern is that pie charts are not always easy to interpret accurately. Discerning small (but sometimes very important) differences based on wedge thickness can be difficult. For this reason, many researchers prefer using bar charts.

Bar Charts

Bar charts are graphs that show how often each category is represented in a distribution, and, in this respect, operate much like histograms. The categories of the variable are listed across the horizontal axis. Each category has a bar and the height of each bar represents the frequency of that category in the data set.[8] There are different variations of bar charts, but we will start with a simple bar chart, which places bars side by side in a single graph. To produce a bar chart of HOMOSEX, perform the following commands:

[8] Bar charts can also be composed so that they are oriented in a horizontal fashion. This is helpful especially if category labels are long, or if there are many bars in the graph.

Graphs
Interactive
Bar
Assign Variables
$pct
(note- this instructs SPSS to list percentages. It is located at the top of the variable list)
HOMOSEX
OK
(If missing values are incorporated into the graph, check *Options*.)

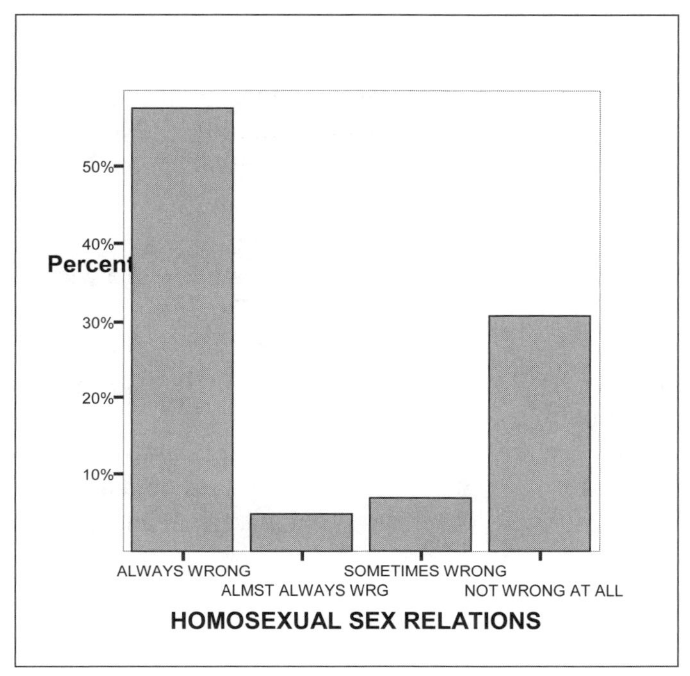

Figure 3.13 Bar Chart of Categorical Variable HOMOSEX

Summary

This chapter introduced some of the essential skills associated with univariate analyses and their application within SPSS and the STATES07 and GSS04 data. Univariate analysis involves understanding the characteristics of a single variable in isolation from other variables. One approach is to find measures of central tendency—the mean, median, and mode. These statistics can be obtained using the *Explore*, *Descriptives,* and *Frequencies* commands. Spread is commonly measured with the standard deviation, the range, and the interquartile range. These can be found using the *Explore* commands. Graphs, however, offer perhaps the most information about distributions. Histograms, pie charts, bar charts, and box plots can be very informative. The following table summarizes the appropriate graphs for scale and categorical variables:

Variable Format	Graph
Scale	Box Plot
	Histogram
Categorical	Pie Chart
	Bar Chart

Figure 3.14 Summary of Graph Types

Once researchers have developed a foundation in the skills of understanding individual variables, they are well positioned to perform accurate and informative analyses of the relationships between variables.

Key Terms

25th Percentile
75th Percentile
Bar chart
Bell-shaped distribution
Bimodal distribution
Box plot
Class
Data cleaning
Distribution
Five number summary
Frequency distribution table
Histogram
Interactive chart editor
Interquartile range
Maximum
Mean
Measures of central tendency
Measures of spread
Median
Minimum
Mode
Negatively skewed distribution
Outlier
Pie chart
Positively skewed distribution
Range
Sorting data
Standard deviation
Univariate analysis

Chapter 3 Exercises

Name___________________________________ Date____________________

1. In advance of performing an analysis, make a prediction: What percent of Americans view their marriages as "very happy", "somewhat happy", and "not too happy"? Then use the GSS04 data, run a frequency on HAPMAR (Happiness of Marriage), and also create a publishable pie chart. Describe your findings below and explain the extent to which your hypotheses conformed with the observations. What do these findings suggest about marriage happiness?

	Predicted Percents	Actual Percents
Very Happy	_______	_______
Pretty Happy	_______	_______
Not Too Happy	_______	_______

How close were your predictions? If they differed, what do you think was the reason?

What do these findings suggest about the state of marriage in America?

2. Using the GSS04 data set, construct and print a bar chart displaying how rushed individuals feel (RUSHED). Give an appropriate title to the graph and refine the graph so that it is clearly printed and aesthetically appealing. Exclude missing cases. What does this graph inform us about the extent of pressure on Americans?

   Thinking ahead: What types of forces may cause Americans to feel rushed? Can you find variables in the GSS04 data that offer indicators of this concern? Create some summary statistics of these indicators and describe them.

   Indicator 1:

   Indicator 2:

3. Using the STATES07 data set, perform a case summary of ENS219 (Hazardous Waste Sites on the National Priority List 2006).

   What three states have the greatest number of waste sites?

   What three states have the fewest waste sites?

   Create a box plot to that illustrates the distribution of these sites. What does this boxplot suggest?

4. Using *Explore,* determine the measures of central tendency and spread for CRC186 (Juvenile Arrest Rate per 100,000 for Violent Crime: 2005). Examine the statistics and box plot, and determine which measures of central tendency and spread are most appropriate.

Mean ____________________

Median ____________________

Standard deviation ____________________

Range ____________________

Interquartile Range ____________________

Circle the measure you would use as the measure of central tendency:

Mean Median Mode

Circle the measure you would use as the measure of spread:

Standard deviation Range IQR

Why did you make these selections?

5. Using *Explore,* determine the measures of central tendency and spread for HTH492 (Adult Per Capita Distilled Spirits Consumption 2004). Examine the statistics and box plot, and determine which measures of central tendency and spread are most appropriate.

   Mean ____________________

   Median ____________________

   Standard deviation ____________________

   Range ____________________

   Interquartile Range ____________________

   Circle the measure you would use as the measure of central tendency:

   Mean Median Mode

   Circle the measure you would use as the measure of spread:

   Standard deviation Range IQR

   Why did you make these selections?

6. Using *Explore,* determine the measures of central tendency and spread for DMS483 (Divorce Rate per 1000: 2005). Examine the statistics and box plot and determine which measures of central tendency and spread are most appropriate.

Mean ____________________

Median ____________________

Standard deviation ____________________

Range ____________________

Interquartile Range ____________________

Circle the measure you would use as the measure of central tendency:

Mean Median Mode

Circle the measure you would use as the measure of spread:

Standard deviation Range IQR

Why did you make these selections?

What state is the outlier?

7. Using *Explore,* determine the measures of central tendency and spread for HTH505 (Percent of Adults Who Do Not Exercise: 2005). Examine the statistics and box plot and determine which measures of central tendency and spread are most appropriate.

   Mean ____________________

   Median ____________________

   Standard deviation ____________________

   Range ____________________

   Interquartile Range ____________________

   Circle the measure you would use as the measure of central tendency:

   Mean Median Mode

   Circle the measure you would use as the measure of spread:

   Standard deviation Range IQR

   Why did you make these selections?

Chapter 4
Constructing Variables

Overview

This chapter shows how to reconfigure existing variables to make them better suited to statistical analyses. We outline the processes of recoding and computing new variables and consider ways to redefine and reorganize existing data. The construction of new variables requires good record keeping and, toward that end, we outline how syntax files can be used to track how data are reorganized. We also show how syntax files can be used to construct new variables and illustrate this process by constructing an index.

Why Construct New Variables?

Social science research often uses data from sources such as police departments, hospitals, and schools. These data can be very informative, but commonly require refinement before they can be used to answer the types of research questions posed by social scientists. For example, a police department may provide the number of crimes that have been reported in an area, but the research may be interested in determining the crime rate, which would involve coupling those data with population data.

Even when the social scientists are in full control of data collection procedures, they commonly need to reconfigure variables. For example, a study may ask how many hours a person works. Ultimately the researcher may need to compare full-time workers with part-time workers, which would involve shifting the scale variable (work hours) to a categorical variable (full-time/part-time). SPSS is designed to help researchers make these types of changes to variables, which can involve transforming the information held in a single variable or combining the information held in two or more variables.

Recoding Existing Variables

There are many circumstances in which a researcher may want to **recode** (or reconfigure) an existing variable. One common case is when the existing variables do not directly measure the concept of interest. For example, alcohol consumption is often measured as a scale variable (0,1,2,3,4...drinks/day). However, American culture emphasizes the distinction between abstainers, social drinkers, and alcoholics. How would you designate these categories if you had data on the number of drinks consumed in a typical day? At what point does one become "alcoholic" or a "heavy drinker"? To answer questions such as this, researchers commonly consider both the prior literature on the subject and examine distributions of the data through univariate analyses. Once decisions about the structure of new variables are made, recoding is usually straightforward. But it is remarkably easy to make computational errors, so any recoding of variables requires data management skills and careful checks on variable constructions.

SPSS offers two choices under the recode command: *Into Same Variable* and *Into Different Variables*. The command *Into Same Variable* replaces existing data with new values, but the command *Into Different Variables* adds a new variable to the data set. In almost every situation, we suggest selecting *Into Different Variables*. Because recoding *Into Same Variables* replaces the values in the existing variable, it will irrevocably alter a data set, threatening your access to the original data.[9] Likewise, mistakes in the recoding statements can accidentally and permanently alter the data. For these reasons, it is best to not recode over an existing variable unless you are absolutely confident that the lost information will never be needed.

To illustrate how to recode a numerical variable into a categorical variable, we will use the STATES07 data to create a new variable indicating whether or not a state performed an execution from 1977-2005. Variable CRC77 (Prisoners Executed: 1977 to 2005) is a scale variable with the number of executions ranging from 0 to 355. To change this to a categorical variable, use the *Recode* command:

Transform
Recode Into Different Variables

In the first window select the variable CRC77 and identify the new *Output Variable* as EXECUTE. *Label* the output variable as "State Had One or More Executions 1977 to 2005." After entering information, click *Change*. You should duplicate Figure 4.1.

[9] One circumstance in which recoding into the same variable is a good practice is when intermediary steps are used to construct a final version of a variable. In those instances, a new variable is created from an old variable using *Into Different Variable*. But then as the new variable is modified through a series of recoding procedures, values are replaced within using *Into Same Variable*. As a result, the original variable remains unchanged, and the data set remains uncluttered with intermediary variables used to construct the final variable. We illustrate this later in this chapter.

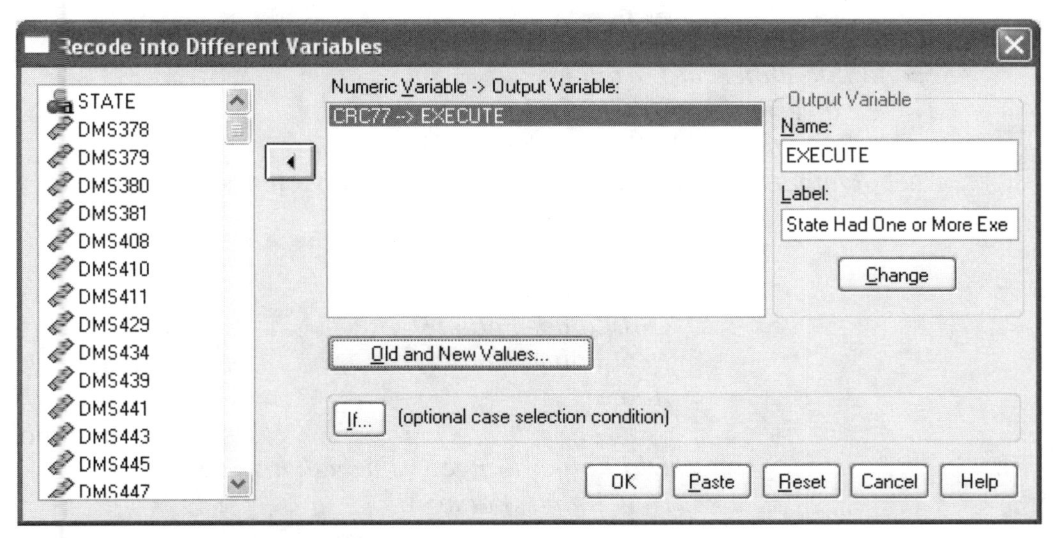

Figure 4.1 First Recode Window

Figure 4.2 Recode Old and New Values

Select *Old and New Values* to reveal another window as shown in Figure 4.2. The left side indicates the values of the old variable, CRC77. The right side indicates the corresponding values of the new variable, EXECUTE. States with no executions (CRC77=0) will be coded as 0 and states with executions (CRC77 ≥ 1) will be coded as 1 in variable EXECUTE. Note that values can be replaced one at a time, or a range of old values can be recoded as a single new value. After entering the old value on the left side and the new value on the right side, click *Add* to record that pair of corresponding values. After all of the new values have been added, click *Continue*, then *OK*.

The whole process is as follows:

Transform
Recode Into Different Variables
Numeric variable: CRC77
Output variable Name: EXECUTE
Output variable Label: State Had One or More Executions 1977 to 2005
Change
Old and New Values
Old Value: Value: 0
New Value: Value: 0
Add

Old Value: Range: 1 *through highest*
New Value: Value: 1
Add

Old Value: System- or User-Missing
New Value: System Missing
Continue
OK

It is important to understand how SPSS deals with missing values. A value that is coded as missing in the original variable will not automatically be coded as missing in the new variable. It will be grouped in with any ranges you specify. Therefore, if the original variable has any values coded as missing, always select *System or User Missing* values in the old variable to be recoded as *System Missing* in the new variable. You can also specify these values in the *Variable View* window.

We strongly recommend that you carefully compare data on a case by case basis to make sure the new variable reflects your intents after each variable recode. An easy way to do this is to generate a report that lists all of the cases, pairing cases of the old and new variables side by side. There is usually no need to check every single case, just scan the data for errors and consistency. To create the output, run the following command:

Analyze
Reports
Case Summaries
Variables: CRC77
EXECUTE
OK

To track new recoded variables, it is also important to label both the variable and its values. The values 0 and 1 can be labeled "no" and "yes" in the *Variable View* spreadsheet, as outlined in Chapter 2. To do this, go to the *Variable View* window, add value labels for the new variable EXECUTE. After constructing and labeling the variable, use the *Frequencies* command to examine the distribution and compare it to the distribution of the original variable. The frequency should indicate that 64.7% of states performed at least one execution from 1977-2005.

Recoding data makes alternative approaches to analyzing relationships possible. Suppose, for example, that we wanted to keep the number of executions in the analysis, but had concerns that Texas (an outlier with 355 executions) was swaying the results. Some types of analyses are more reliable if an outlier is excluded from the analysis and treated as a separate unique observation. How could we create a new variable to help us deal with this situation?

Conditional transformations offer a useful approach to recoding Texas. This is done using the *IF* function in the *recode* procedures. To signify the transformation of the observation for Texas, we will use its STATEID (44), a unique identifying number for this observation[10]. We will do this in two steps, first creating a new variable that is identical to CRC77, and then performing a conditional transformation of that variable.

Transform
- *Recode Into Different Variables*
  - *Numeric variable:* CRC77
  - *Output variable Name:* EXECUTE2
  - *Output variable Label:* Number of Executions 1977 to 2005, Texas Recoded Missing
  - *Change*
  - *Old and New Values*
    - *Old Value: Range, value through HIGHEST:* 0
    - *New Value: Copy old values*
  - *Old Value: System- or User-Missing*
  - *New Value: System Missing*
    - *Continue*
  - *OK*
- *Transform*
  - *Recode Into Same Variables*
    - *Variables:* EXECUTE2
  - *IF…*
    - *Include if case satisfies condition:*
      - STATEID=44
    - *Continue*
  - *Old and New Values*
    - *Old Values: Range, value through HIGHEST:* 0
    - *New Value: System-Missing*
  - *Add*
  - *Continue*
- *OK*

[10] Most data sets of individuals, organizations, or places have a unique identification number, such as RESPONDENT ID, that helps locate the specific observation within the data set. In this case, we simply assigned STATEID in the STATES07 data set by assigning progressive numbers to the states in their alphabetic order.

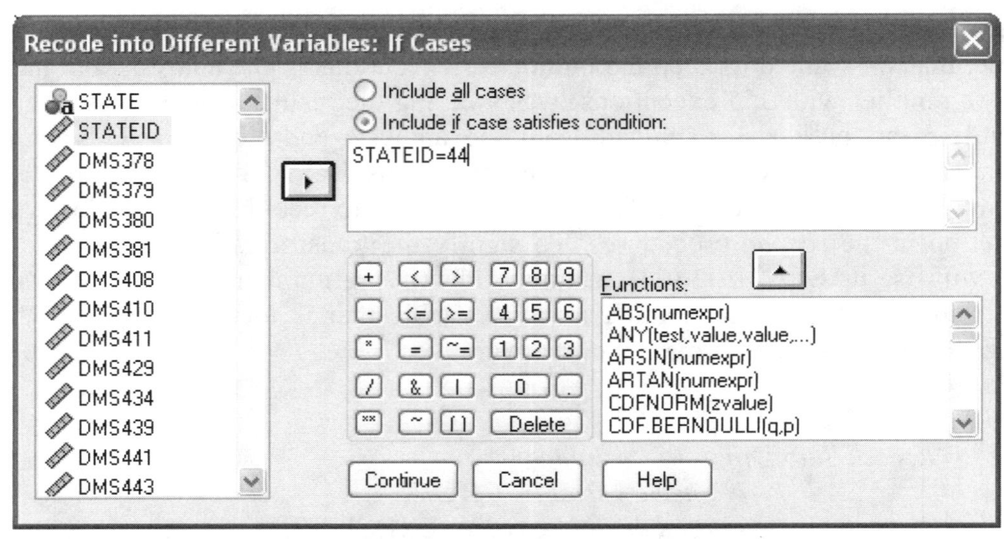

Figure 4.3 The "If" Conditional Commands Dialogue Box

After constructing this variable, remember to check to see that your commands operate as intended using *Reports*.

Analyze
  Reports
    Case Summaries
      Variables: STATEID
        STATE
        CRC77
        EXECUTE2
      OK

Is Texas coded as missing and are all other values the same as in the original data?

Computing New Variables

In some cases, the concern is not simply to restructure the information in an existing variable, but to combine the information in two (or more) different variables. Notice that many of the variables in the STATES07 data are in the form of **rates** (such as the number of violent crimes occurring per 100,000 people). These variables required combining the incidence figures with population figures. In constructing the data for this book, we tried to anticipate many of your analytic needs, but here (and in future projects) you may need to rework the data by computing new values. Below we will illustrate the process of constructing a new variable by computing the percentage of all crimes that are violent crimes.

Calculating the percentage of crimes that are violent crimes involves combining the information from variables CRS26 (Crimes: 2005) and CRS30 (Violent Crimes: 2005). Use the *Compute* command to calculate a new variable "VICRIMPCT." You can also *label* the new variable "Violent Crimes as Percentage of All Crimes 2005" from the *Compute* window.

Transform
Compute
Target Variable = VICRIMPCT
Numeric Expression = (CRS30 /CRS26)*100
Type&Label
Label: Violent Crimes as Percentage of All Crimes: 2005
Continue
OK

You can enter the information in the *Numeric Expression* box either by typing in the numeric expression (above) or by using the mouse to choose the variables from the list at the left and the mathematical signs from the calculator keyboard. Remember to document how you computed VICRIMPCT by entering a *Variable Label* right away. We suggest "Violent Crimes as Percentage of All Crimes: 2005." Figure 4.4 shows how to generate the new *Target* variable name VICRIMPCT using the *Compute* command.

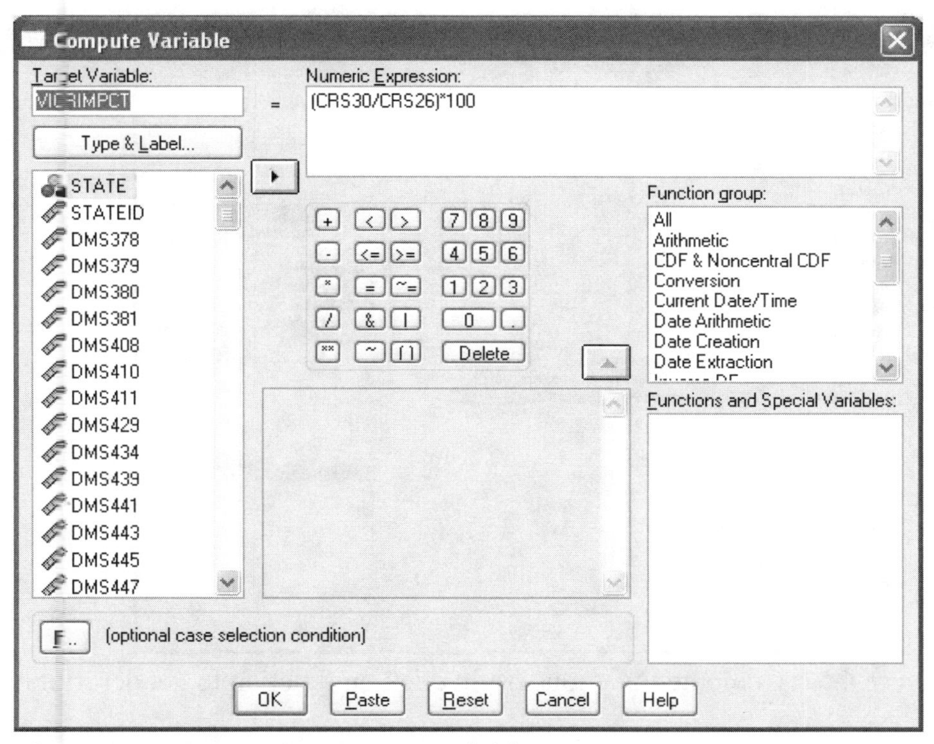

Figure 4.4 Computing a New Variable

The speed and ease with which SPSS creates new variables can be seductive. Even when researchers are as careful as possible, mistakes inevitably occur. For this reason, before making any conclusions about violent crime as it compares to other crimes, it is extremely important to inspect the new variable to confirm that it was accurately calculated. This process is part of **cleaning data**—ridding the data of errors. The types of univariate analyses outlined in the previous chapter offer multiple ways of checking if new variables are accurately computed. For instance, *Frequencies* or *Descriptives* can be used to quickly check that all values fall within possible ranges, a procedure termed **possible code cleaning**. In this case, no percentage should be higher than 100% or lower than 0%. Another approach is to engage in **contingency cleaning**—comparing the constructed variables to other existing variables. You can also use intuition to see if variables have realistic values and if measures of central tendency (means, medians, or modes) and spread (standard deviations) appear reasonable. Knowing that violent crimes occur far less frequently than property crimes, any values that indicate violent crimes constitute the majority of crime would be surprising to observe.

The most common computations involve adding, multiplying, dividing, and subtracting variables from one another. Yet researchers may sometimes want to perform other types of computations, such as changing the structure of date variables, calculating elapsed time from dates, doing logarithmic transformations, or a wide range of other functions. To gain a sense of the range of procedures available, look through the functions listed in the *Compute* window. For a description of each function, simply highlight it using the left mouse button, then click on the right mouse button. A text box explains the type of operation that this function performs, as shown in Figure 4.5.

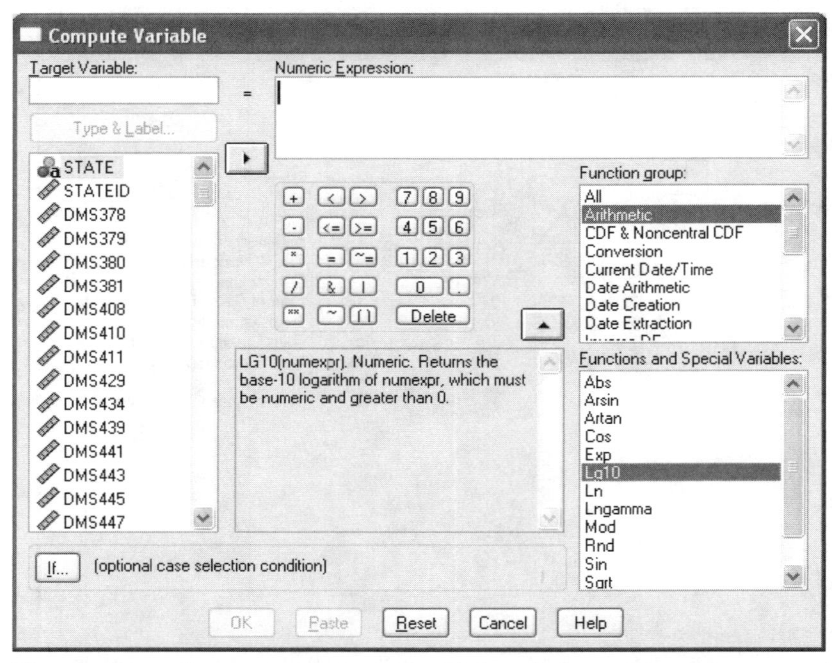

Figure 4.5 Compute Functions (right click on function to get descriptions)

Recording Computations Using Syntax

Computing new variables and using them in data analysis requires considerable record keeping and data management skills. For this reason, we recommend that researchers keep extensive notes on the ways they computed their variables. One way of doing so is to paste computations into syntax files using the *Paste* command within the *Compute* window. **Syntax files** contain procedures written in the SPSS command language that, when submitted, cause the SPSS processor to perform statistical operations. Syntax files are an alternative, as well as a supplement, to the dialog windows.

To illustrate the process and value of creating syntax records, we will construct an **index variable**, representing a composite indicator of unhealthy lifestyles. An **index** is a single score that summarizes responses to two or more variables. Indices are often used when researchers measure an underlying concept, such as depression or work commitment, that cannot be reliably measured with a single indicator. In such a situation, one approach is to combine multiple variables into a single indicator, which makes analysis less repetitious and statistical summaries meaningful. For example, a researcher who is interested in understanding attitudes toward abortion may find it advantageous to ask a number of questions, as most people will fall within a continuum between believing that abortion is wrong in every circumstance and believing that abortion is permissible in any situation.

To illustrate index construction and how syntax files operate, we will combine together two variables, the percentage of adults who smoke (variable HTH495) and the percent of adults who do not exercise (HTH504), to create a new variable UNHEALTHYLS (Unhealthy Lifestyle Index). We will do this with the compute function, but in this case before saying *OK*, we will ask SPSS to paste the commands into a syntax window. Replicate Figure 4.6, and then click *Paste*, and you will see a new window open that looks like Figure 4.7.

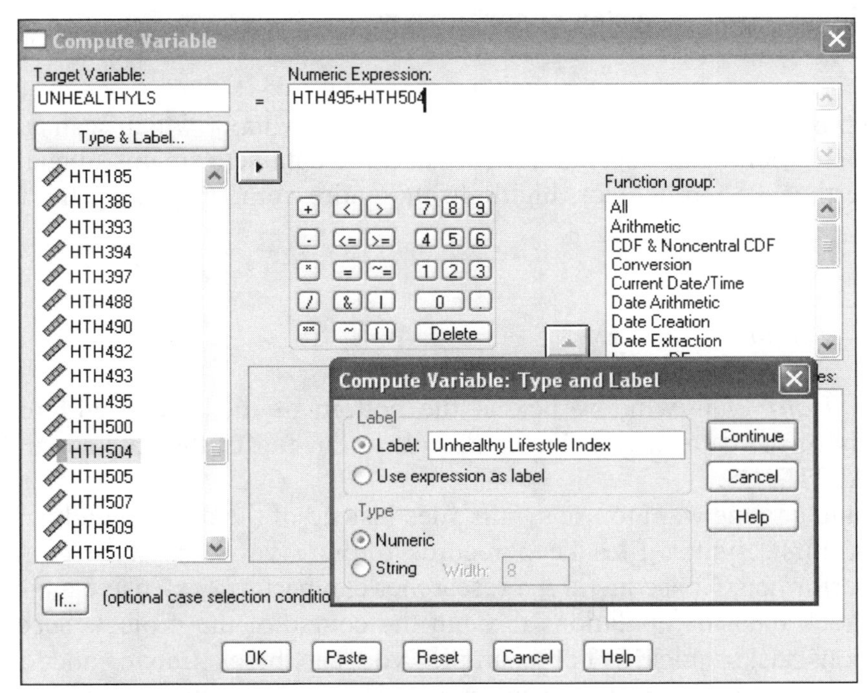

Figure 4.6 Computing an Index by Adding Two Variables

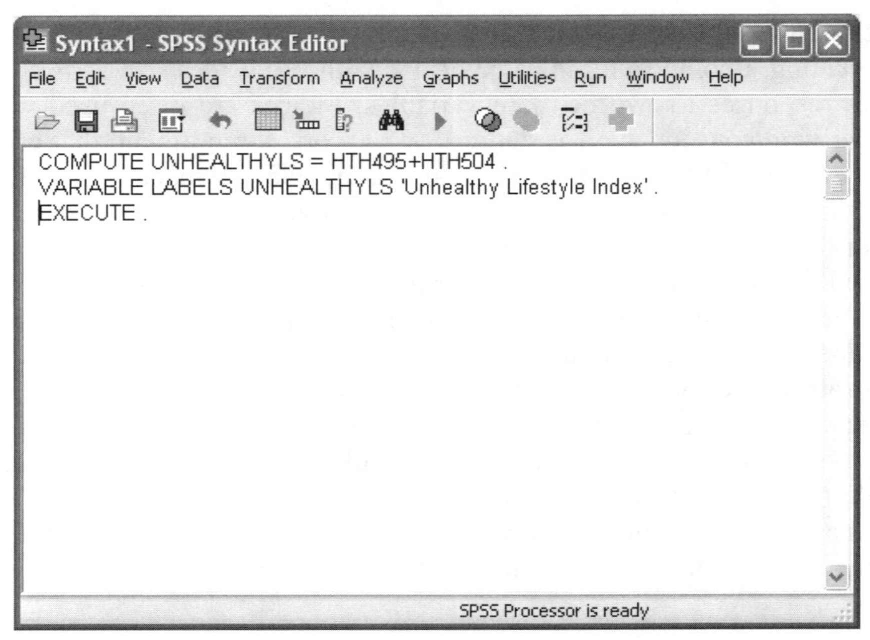

Figure 4.7 A Syntax Editor Window

The command has now been summarized in syntax, which can make SPSS run procedures. But this command has not yet run, and the data set remains unchanged. To submit the command to the SPSS processor, highlight the entire command using the mouse, and then select:

Run
Selection

Clicking on the *Data View* window box at the bottom of the screen will return you to the spreadsheet array of the data, where you can scroll to the end of the data set and see that SPSS created UNHEALTHYLS.

Why would anyone want to use syntax files instead of the dialog windows? There are two general reasons. First, syntax files keep records of how variables were computed and how analyses were performed. Consequently, a researcher can perform extensive analyses and use the syntax files to track data manipulations through the course of the project. Second, syntax files perform repetitious tasks quickly. For example, you can block, move, and copy portions of syntax files just as in word processing programs. Using the syntax to compute many similar variables can enhance the efficiency of constructing, or even analyzing, variables. Once users become familiar with SPSS, many largely dispense with the dialogue boxes and do almost everything through syntax files. The *Help* menu in SPSS shows how to access syntax commands directly, without pasting. For example, syntax commands can label variables (as well as values) directly with:

```
VARIABLE LABELS UNHEALTHYLS 'Unhealthy Lifestyle Index' .
```

The following command saves a syntax file:

File

Save As: UNHEALTHY INDEX COMPUTATION

It can also be reopened and subsequently modified using the command:

File

Open

Syntax

File name: UNHEALTHY INDEX COMPUTATION.SPS

Just as managing data is a concern, so is managing syntax files. We suggest creating two general types of syntax files, **preparation files** and **analysis files**. Preparation syntax files contain commands for constructing and modifying variables. Analysis syntax files contain commands for data analyses for reports. This is very helpful when an analysis needs to be replicated later or described accurately in the report.

As you prepare both types of files, it is often helpful to insert notes within the syntax files to track how and why you manipulated data. These notes are text that SPSS ignores when running commands. To insert a note within an SPSS syntax file, use the asterisk (*) character at the beginning and end of the note and separate the note with at least one blank line from the next command within the syntax file. For example:

```
*The following computes the unhealthy lifestyle index*

COMPUTE UNHEALTHYLS = HTH495+HTH504.
VARIABLE LABELS UNHEALTHYLS 'Unhealthy Lifestyle Index'.
EXECUTE.
```

Combining Recode and Compute Functions: An Illustration

In this final illustration, we will put together many of the recoding processes into a single set of interconnected operations. We will use the GSS04 data set to construct an index measuring self esteem. Our goal is to combine the information contained in five variables into a single composite measure of self esteem:

SATSELF "On the whole I am satisfied with myself"
AFAILURE "I am inclined to feel I am a failure"
OFWORTH "I am a person of worth at least equal to others"
NOGOOD "At times I think I am no good at all"
SLFRSPCT "I wish I could have more respect for myself."

Each of these variables is similarly phrased, with respondents reporting, on a four point scale, whether they strongly disagree (1) to strongly agree (4) with the statement in the variable label. But here is our first challenge. Strongly agreeing to three of the questions (AFAILURE, NOGOOD, SLFRSPCT) indicates lower self esteem, but strongly agreeing to the other questions (SATSELF, OFWORTH) indicates high self esteem (to observe the values, run a *Frequencies* on these variables). To fix this problem, we will need to reverse code the variables that accord high values with beliefs that correspond with low self esteem. You have already learned the skills to do this. Figures 4.8 and 4.9 show the process of inverting the values in AFAILURE so that high scores indicate higher levels of self esteem. Once you do this, repeat the process with variables NOGOOD and SLFRSPCT).

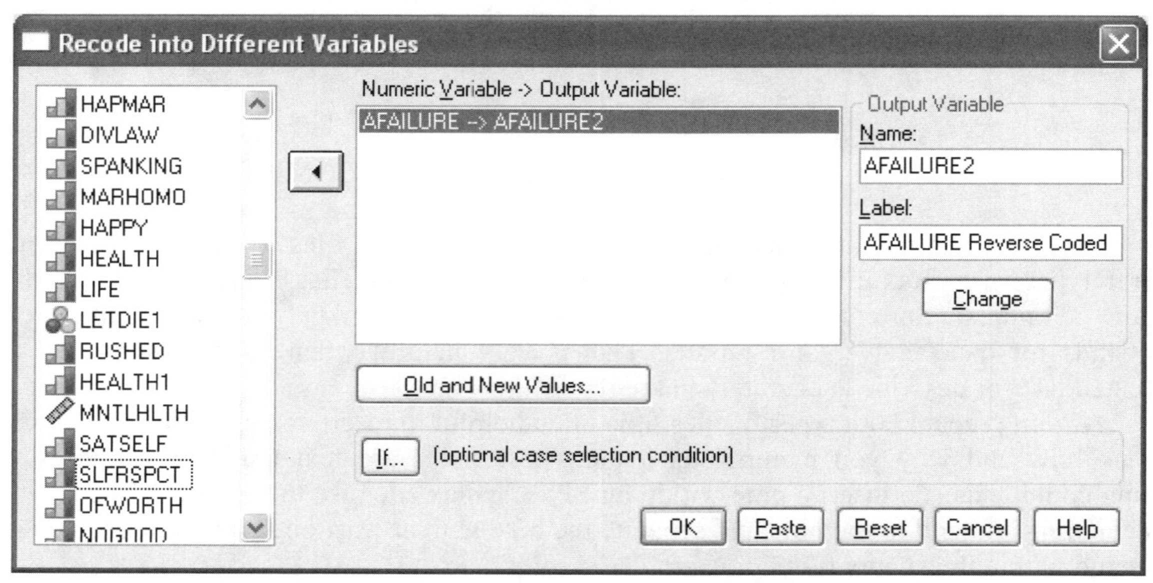

Figure 4.8 Reverse Coding a Likert Type (Agree-Disagree) Variable

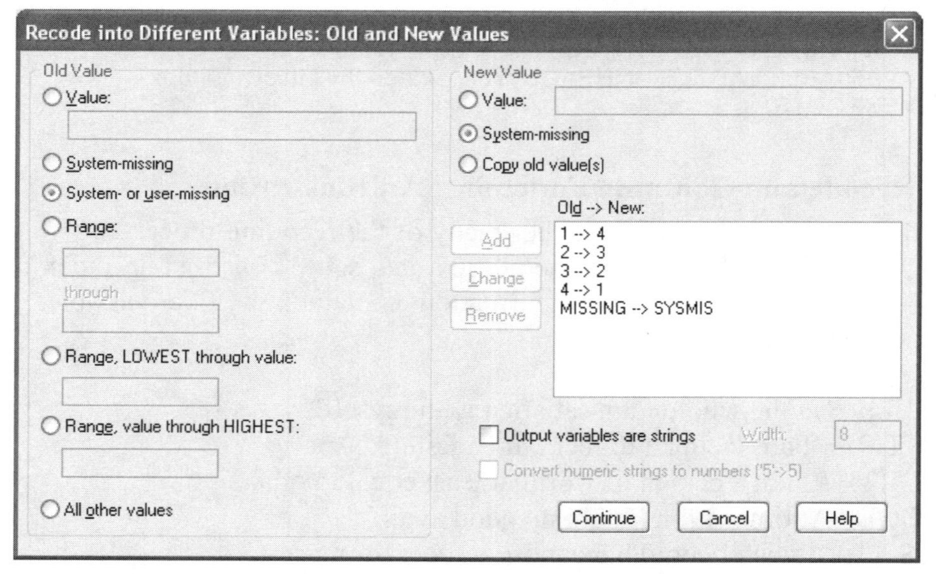

Figure 4.9 Reverse Coding a Likert Type Variable (Continued)

We are now ready to create the self esteem index. This new variable, SELFESTEEM, will combine five variables. While we could simply add the values together, or compute an average, if we did so, we would lose a number of observations because of missing responses to individual items used to compute the index. Therefore, we will take advantage of the *MEAN* function (in this case *MEAN.4*) that will create an average value for all of the variables, so long as there are 4 or more valid observations in each case. Replicate Figure 4.10. Note - to display the *MEAN* function in the lower right dialog box, first click *All* in the *Function and Special Variables* (middle right dialog box).

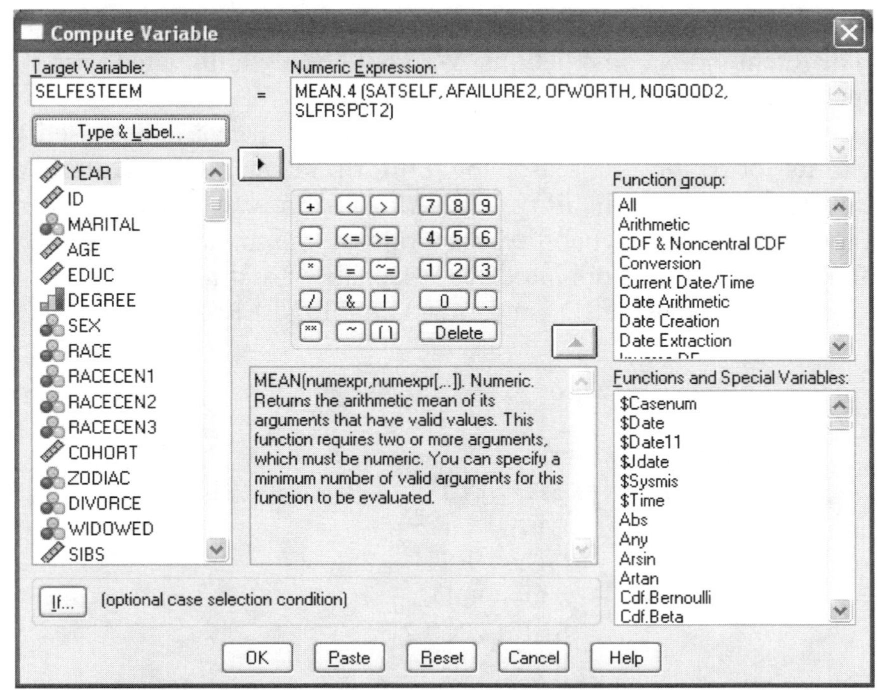

Figure 4.10 Computing a mean of five variables, permitting one missing value

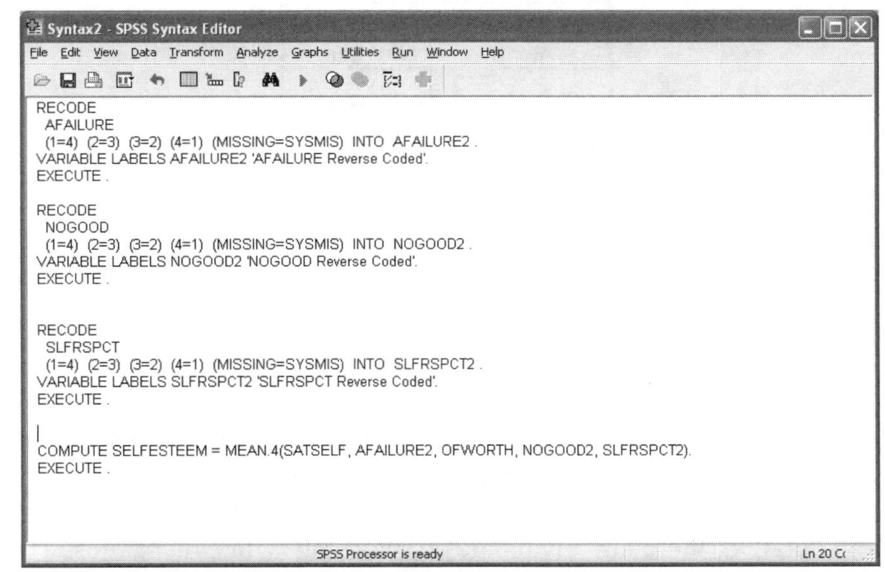

Figure 4.11 A syntax file documenting recoding and computations in the creation of the SELFESTEEM index

Although we created this index using a dialog window, we could have also created it using a syntax file. Recall that clicking *Paste* in this dialog window will reproduce the commands in syntax format. In this case, the syntax would be compiled into a file that looks like Figure 4.11. Aside from documenting operations, syntax offers have an additional advantage, in that repetitious operations can be performed quickly. For example, once we did the recode for variable AFAILURE, we simply cut and pasted those operations and replaced the variable name with NOGOOD and SLFRSPCT in subsequent commands. Additionally, suppose we wanted to experiment with different index constructions. We could easily eliminate (or add) variables and then rerun the operation from the *Syntax* window.

An index should indicate scale reliability, and that the variables used to construct the scale contribute to its formation in a reasonably uniform manner. The **Alpha Coefficient** offers useful information to test scale reliability, The Alpha score can range between 0 and 1, with higher numbers indicating higher reliability. As a rule of thumb, an Alpha score of .70 or higher on an index of four or more indicators indicates reasonably good reliability. The following SPSS commands will yield the Alpha Coefficient for the variable SELFESTEEM:

Analyze
Scale
Reliability Analysis
Items = SATSELF
AFAILURE2
OFWORTH
NOGOOD2
SLFRSPCT2
OK

Reliability Statistics

Cronbach's Alpha	N of Items
.712	5

Figure 4.12 Reliability Analysis Output

The output replicated in Figure 4.12 reveals an Alpha = .712, which indicates reasonably good scale reliability. The various indicators tend to cluster together statistically, which gives us some measure of confidence in the application of the index to the studying of self esteem.

Summary

This chapter demonstrated the two primary ways of reorganizing data within SPSS, computing new variables and recoding existing variables. As new variables are created, they need to be checked for accurate construction. As familiarity with SPSS increases, and as analysis becomes more complex and detailed, there are advantages to working with syntax files. These files can be used to perform repetitious compute or recode procedures. They can also be a valuable means of recording how variables have been constructed and refined. One such circumstance would be in the construction of an index, which often requires many related commands before the final index score is created.

Key Terms

- Alpha Coefficient
- Analysis files
- Cleaning data
- Computing variables
- Contingency cleaning
- Index
- Possible code cleaning
- Preparation files
- Recode
- Syntax files

Chapter 4 Exercises

Name ______________________________ Date ____________

1. Using the STATES07 data, generate a new variable MURDPCT and label it "Murders as % of Violent Crimes" using *Compute.* Use the following formula to generate the new variable.

   (CRS34 / CRS30)*100
   (Murders 2005 divided by Violent Crimes 2005) multiplied by 100

   What is the mean of this new variable?

   ____

   Which state has the highest percentage of violent crimes that are murders?

   ____

   Which state has the lowest percentage of violent crimes that are murders?

   ____

What is the percentage of violent crimes that are murders in your state?

____

2. Using the STATES07 data, *compute* a new variable BURGPCT that determines the percentage of crimes that are burglaries. Use variables CRS26 (Crimes 2005) and CRS47 (Burglaries 2005) to compute the new variable.

   What is the mean of BURGPCT?

   ____

   Which state has the highest percentage of crimes that are burglaries?

   ____

   Which state has the lowest percentage of crimes that are burglaries?

   ____

   What percentage of crimes are burglaries in your state?

   ____

3. Using the GSS04 data, transform (*Into Different Variables*) the variable SIBS (number of siblings) into a new variable, FAMSIZE (number of children in a family) that represents the number of children in the family the respondent was raised in. Note that as SIBS represented the number of brothers and sisters the respondent had and FAMSIZE will be one greater than that number, as the respondent is included as a child in the family. Recode FAMSIZE into the following categories:

    1 Only Children (0 siblings, total of 1 child in the family)
    2 Small Families (1-2 siblings, total of 2-3 children)
    3 Large Families (3-4 siblings, total of 4-5 children)
    4 Very Large Families (5 or more siblings, total of 6 or more children)

    Label the variable "Family Size Categories" and use Value Labels to enter the text above.

    Generate and print a Case Summaries to compare SIBS to FAMSIZE. Did it come out correctly?

    Analyze
    Report
    Case Summaries
    SIBS
    FAMSIZE
    Limit Cases to First 25

    Run a frequency on FAMSIZE and a descriptive of SIBS and answer the following questions:

    What is the most typical number of children in a family? ____________________

    What percentage of respondents were only children? ____________________

    What percentage of respondents came from very large families? ____________________

    What variable do you think would be better to use if one were to compare, for instance, how family size affects children's later socio-economic achievement? Would it be better to use the scale variable SIBS or group families into categories as in the variable FAMSIZE? Explain your reasoning.

4. In this exercise you will work on a conditional transformation. In the GSS04 data, the variable NUMWOMEN indicates the number of female sex partners since age 18, and NUMMEN indicates the number of male sex partners since age 18. But these variables have different implications depending on whether the respondent is male or female, as men are more likely to have female sex partners and women are more likely to have male sex partners. In this exercise, you will generate two new variables.

   NUMOPPSEX - the number of sex partners from the opposite sex
   NUMSAMESEX - indicates the number of sex partners from the same sex

   To do this, use the *Recode Into New Variable* command to create two new variables. Use the optional case selection condition *IF* to specify the variables SEX=0 (the value for women) and SEX=1 (the value for men). For the *Numeric Variables* (these are the input variables) use NUMWOMEN and NUMMEN. Specify NUMOPPSEX and NUMSAMESEX as the *Output Variables*. Remember to recode missing values as well. Check your results by generating a report using the following command

   Analyze
   Report
   Case Summaries
   SEX
   NUMMEN
   NUMOPPSEX
   NUMWOMEN
   NUMSAMESEX
   Limit Cases to First 25

   If you did all of the recoding correctly, your output should look like Figure 4.15. Demonstrate your success by printing this report or do as your instructor suggests.

   NOTE- This exercise may take some head scratching. If you get totally stumped, we offer a hint at the end of this chapter. Use this only as a last resort. Likely you will make a few mistakes along the way, but you should be able to figure this out with a bit of trial and error.

Case Summaries[a]

	SEX RESPONDENTS SEX	NUMMEN NUMBER OF MALE SEX PARTNERS SINCE 18	NUMOPPSEX NUMBER OPPOSITE SEX PARTNERS	NUMWOMEN NUMBER OF FEMALE SEX PARTNERS SINCE 18	NUMSAMESEX NUMBER SAME SEX PARTNERS
1	0 FEMALE	2	2.00	0	.00
2	0 FEMALE	3	3.00	0	.00
3	1 MALE	0	36.00	36	.00
4	1 MALE	-1 NAP	.	-1 NAP	.
5	0 FEMALE	6	6.00	0	.00
6	1 MALE	0	18.00	18	.00
7	1 MALE	0	65.00	65	.00
8	1 MALE	999 NA	.	999 NA	.
9	1 MALE	0	4.00	4	.00
10	1 MALE	0	20.00	20	.00
11	1 MALE	-1 NAP	.	-1 NAP	.
12	0 FEMALE	-1 NAP	.	-1 NAP	.
13	1 MALE	0	99.00	99	.00
14	1 MALE	-1 NAP	.	-1 NAP	.
15	1 MALE	0	.	995 MANY,LOTS	.00
16	1 MALE	0	.	995 MANY,LOTS	.00
17	0 FEMALE	1	1.00	0	.00
18	1 MALE	0	.	995 MANY,LOTS	.00
19	0 FEMALE	1	1.00	0	.00
20	1 MALE	0	30.00	30	.00
21	0 FEMALE	2	2.00	0	.00
22	0 FEMALE	1	1.00	0	.00
23	1 MALE	0	98.00	98	.00
24	1 MALE	-1 NAP	.	-1 NAP	.
25	1 MALE	0	2.00	2	.00
Total N	25	19	16	16	19

a. Limited to first 25 cases.

Figure 4.15 A Report of Variables Generated from Conditional Commands

5. Now that you have constructed these variables, run a frequency and descriptive of NUMOPPSEX. Briefly describe what you observe in respect to the social norms of engaging in relations with members of the opposite sex. How would you describe these norms in the United States?

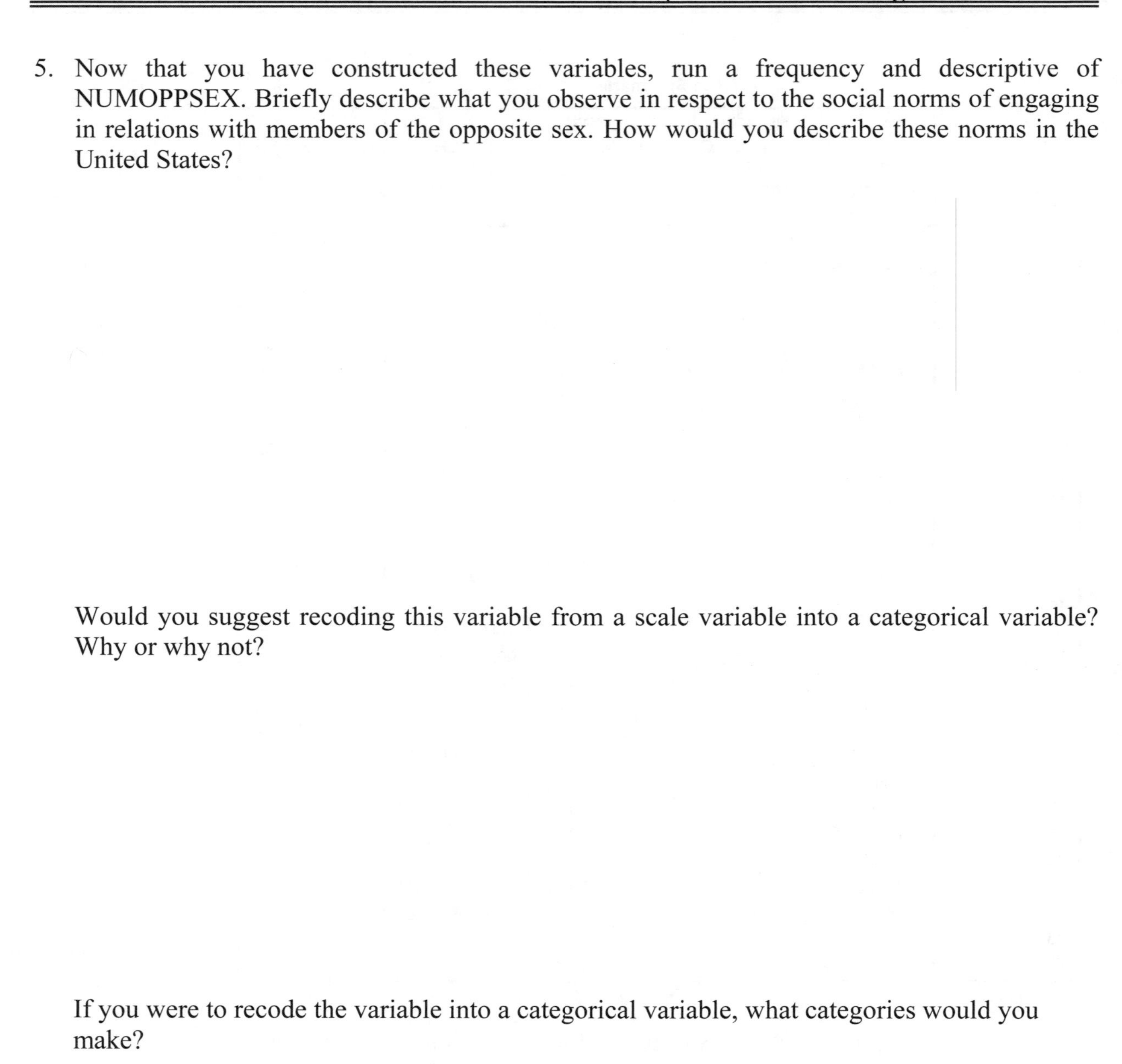

Would you suggest recoding this variable from a scale variable into a categorical variable? Why or why not?

If you were to recode the variable into a categorical variable, what categories would you make?

6. Run a frequency and descriptive of NUMSAMESEX. Briefly describe what you observe in respect to the number of same sex partners with whom people have had relations. How would you describe the norms of same sex engagement in the United States based on these data?

   Would you suggest recoding this variable from a scale variable into a categorical variable? Why or why not?

   If you were to recode the variable into a categorical variable, what categories would you make?

7. Notice that there are a number of people who are not coded into the new variables NUMOPPSEX and NUMSAMESEX and they are treated as missing values. But there is some valuable information available on many individuals in the original variables NUMWOMEN and NUMMEN.

   Those who were originally coded as "NAP" (Not applicable) were never asked the question concerning sexual partnering. However, a number of respondents said that they had "Some 1+" "MANY" or "LOTS" of sex partners. Here is a conundrum, what is the best way to treat these observations? Is there any way to include these missing cases in the new variables? Should you assign a numerical value to these cases, and, if so, what should that number be? If you keep them as missing, are you then leaving out of the study some of the most important observations? What should a good analyst do?

Exercise 4 Hint. This will get you moving. This is the coding of women and their number of opposite-sex partners. But to fully create the new variable NUMSAMESEX and NUMOPPSEX, you have to think about how to specify the conditional statement *IF* (SEX = 0 or SEX = 1) and the *Numeric Variable* (input) and the *Output Variable* for these other combinations:

Women's same sex partners
Men's opposite sex partners
Men's same sex partners.

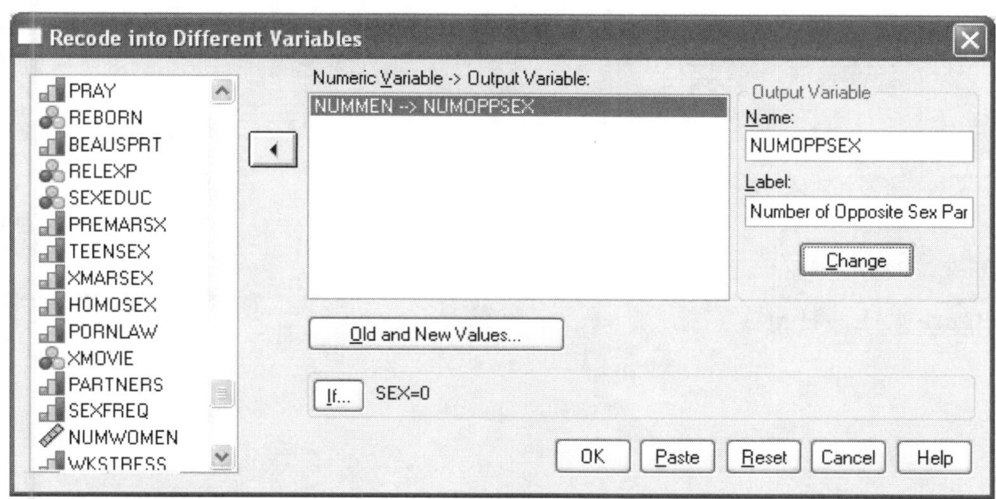

Figure 4.16 Hint 1

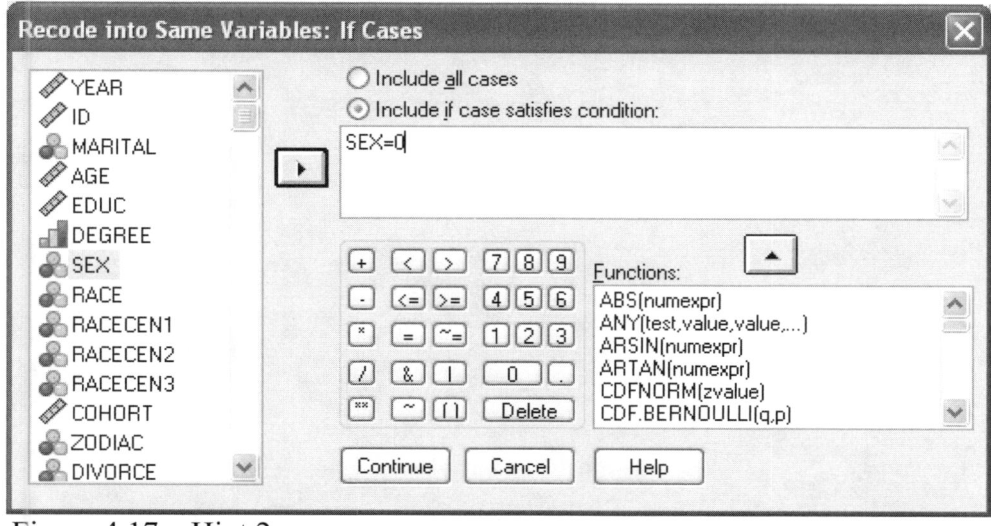

Figure 4.17 Hint 2

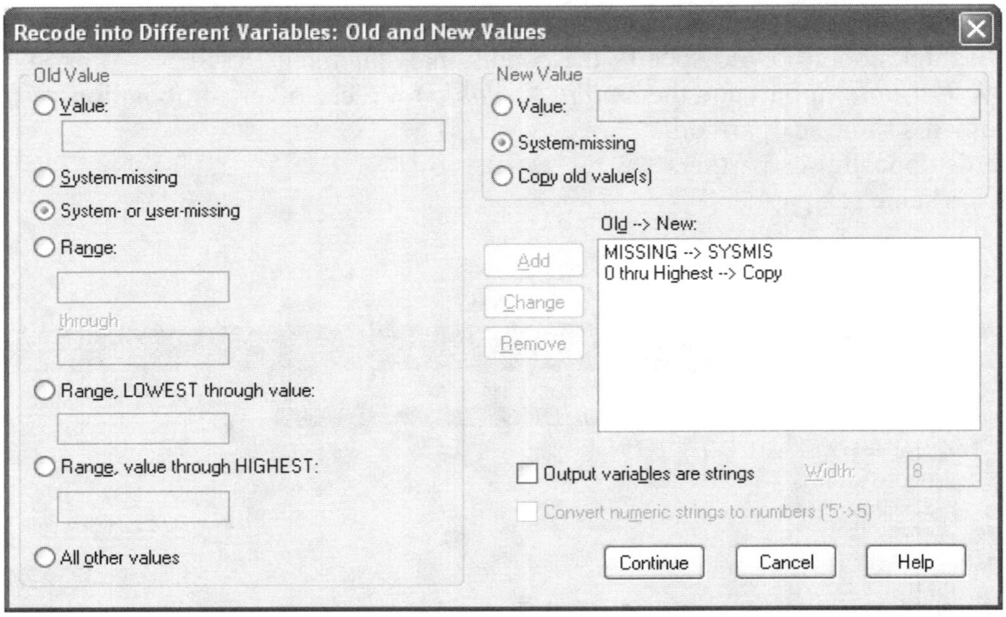

Recode into Different Variables: Old and New Values

Old Value
Value:
System-missing
System- or user-missing
Range:
through
Range, LOWEST through value:
Range, value through HIGHEST:
All other values

New Value
Value:
System-missing
Copy old value(s)

Old --> New:
MISSING --> SYSMIS
0 thru Highest --> Copy

Add
Change
Remove

Output variables are strings Width: 8
Convert numeric strings to numbers ('5'->5)

Continue Cancel Help

Figure 4.18 Hint 3

Chapter 5
Assessing Association through Bivariate Analysis

Overview

Many of the most interesting questions addressed by social scientists examine relationships between two variables. For example:

Is the death penalty associated with lower crime rates?
Does school funding relate to students' educational success?
Are religious people more politically conservative?

Each of these questions asks if the values of one variable (such as whether or not a state has the death penalty) are associated with the values of another variable (such as crime rates).

Our introduction to bivariate (two variable) analysis begins with an explanation of **significance tests**, methods for determining if a relationship observed in the sample is the result of chance and the likelihood that this same relationship also exists in the population. We then explain how significance tests apply to bivariate associations. This chapter focuses on two types of relationships - cross tabulations and correlations. **Cross tabulations** measure the association between two categorical variables, such as ethnicity and religion. **Correlations** measure the association between two scale variables, such as income and age. We also show ways of graphing bivariate associations, which are useful for exploring variable distributions affect the relationships and for disseminating findings in reports and presentations. In Chapter 6, we extend the discussion of bivariate analyses to methods of testing and graphing associations between scale and categorical variables.

Why Do We Need Significance Tests?

In an ideal world, data analysts would have access to every possible observation, such as knowing the behavior and perspectives of every person in the world. Of course the reality is that we have only a small portion of potential observations represented in any study. For instance, in this book you are asked to gauge behaviors and attitudes in the United States (a country with over 300 million residents) on the basis of a survey that has fewer than 3,000 participants. This presents a challenge. How do we know that relationships observed in this small sample also exist in the much larger population?

Social scientists use significance tests to determine the likelihood that a relationship between two or more variables is due to chance occurrence. These tests determine if a relationship is **statistically significant**, which means that an observed pattern would likely continue to exist if we took another sample from the entire population, and that the pattern would be evident if we actually were able to study the whole population.[11]

As we discussed in Chapter 1, the need for significance tests comes from using samples to make inferences about populations. Significance tests require **random samples**, in which every element from the population has an equal chance of being selected for inclusion in the study. Owing to the laws of probability, data from random samples are *generally* similar to the data in the whole population, but every random sample is not exactly the same. For example, one in twelve Americans is African-American. Thus, we would expect that a random sample of 1,000 Americans would include approximately 120 African Americans. But if we pulled many different samples, that representation would fluctuate, with sometimes an over-representation and sometimes an under-representation. Were it possible to take *many* random samples, the distributions would look very much like the population. But since data collection almost always involves taking a single sample, that sample may, just by random chance, look a bit different from the population, and in turn influence findings in ways that do not reflect the patterns in the population. Here we describe two examples of how random occurrences can look like real patterns.

Consider this possibility. Suppose we wanted to study men's commitment to housework. A variety of studies show that men tend to have relatively low commitments to housework compared to women, and that men's employment has relatively low effects on the volume of housework they perform. In other words, men who work long hours tend to do about the same amount of housework as men who work short hours – not all that much. In generating a sample for a study, one possibility is that, simply by the luck of the draw, we obtain a disproportionate number of highly successful sensitive new age guys in the study. Analysis of this sample would indicate that the more work men perform outside the home, the greater their commitment to housework. It would also suggest that men's commitment to housework is not all that different than women's. Herein lays the danger. By generating an odd sample, we create odd and untrustworthy findings that do not reflect the characteristics of the population.

An obvious solution would be to draw a new sample and see if the same results hold. After all, what is the likelihood of obtaining a disproportionate number of sensitive new age guys twice in a row? The next time around, chances are we would end up with a more typical sample of men, and maybe even get an over-representation of couch potatoes. In those instances our findings would be negated, and possibly even reversed. But the reality is that most researchers do not have the luxury to pull numerous samples and, because they only have one sample to work with, significance tests are essential to making conclusions that observed relationships are unlikely to be attributed to such odd-ball occurrences in sample characteristics.

[11] As we discuss later, a statistically significant relationship does not necessarily mean that the relationship is strong, reflects causality, or even that the relationship is important.

Here is another way of thinking about the concern of random variation and how it creates the appearance of relationships that in reality do not exist. Our friend, Santosh Venkatesh, at the University of Pennsylvania, uses a unique method to illustrate how chance occurrences often look like they have strong patterns. Santosh divides his classes into two groups of students. One half of his class is assigned the task of fooling him by creating fake documents that illustrate the result of one hundred coin tosses. The other half actually flips a coin 100 times. Thus, in a class of 40 students, he receives 20 documents showing actual coin tosses and 20 documents summarizing sequences of fictitious coin tosses. The students record heads with "H" and tails with "T" (HTHTHHTTHTHTHT…etc.). To his students' amazement, Santosh can distinguish the faked lists from the actual coin flips with remarkable consistency (he is right about 80% of the time). How does he do this? Not with SPSS or even a hand calculator.

Santosh uses a simple principle to guide his sleuthing - his knowledge that students will tend to be insensitive to the degree of variation normally found in random occurrences. For example, students will tend to fake coin tosses by alternating heads and tails with great regularity (HTHTHTTHTHT). In reality (and you can try this), numerous repeats of an event (HTHHHHHTTTH) are commonly observed in long sequences of random events. In any sequence of 100 coin flips, it is highly likely[12] that there will be a sequence of at least 5 heads or tails in a row. Very rarely do students produce faked lists containing such a long sequence. His students also tend to make sure there are an equal number of heads (50) and tails (50) in the series. Although a random event will *tend to* produce equal outcomes, unequal outcomes are actually more likely. For example, there is actually a much greater likelihood of having an unequal number of heads and tails than achieving exactly 50 heads and 50 tails in one sequence of 100 tosses.

Our friend's trick is relevant to the topic of significance tests because it highlights the fact that random events do not necessarily look random. It is common for any set of observations to look like it forms a relationship when, in fact, it is a result of chance occurrence.

Hypotheses and Significance Tests

Patterns in data are caused either by chance or by the existence of real relationships. By "real" we do not mean that one variable causes change in the other variable, only that there is a discernable association in the full population, such as between temperature and sweating or age and mortality. In testing whether or not a relationship is real, a data analyst considers opposing statements, called hypotheses. The **null hypothesis** says that the relationship is due to chance. The **alternative hypothesis** says that the relationship is real. The data analyst's job is to use probabilities to judge which hypothesis is more strongly supported. This can be tricky business. Analysts never know which result will occur, so they set up analytic tests.

In their training, data analysts are coached to adopt a conservative disposition when viewing statistical output. Their job is to operate on the assumption that relationships *do not exist* (the null hypothesis is supported) unless significance tests overwhelmingly indicate otherwise. This is helpful because it forces analysts to be skeptical in making claims and to set a high threshold for relationships to be taken seriously. In other words, they want a very strong indication that chance is not at play in the findings, and are unwilling to accept the appearance of a relationship that may not in fact exist.

[12] The probability is .80. Probability is defined on the next page.

It is important to also recognize that analysts structure their statistical models in respect to both null and alternative hypotheses. It is reasonable to approach any data set with the expectation that certain types of relationships found in past studies and theories will be evident in the current study. These expectations guide which variables to include in the statistical models and how to structure these models.

Significance Levels

Significance tests rely on the **significance level**, the probability that chance explains patterns in the data.[13] A significance level with a high value supports the null hypothesis: chance is highly likely to explain the pattern in the data. Conversely, a low significance level supports the alternative hypothesis: chance is unlikely to have caused the pattern in the data and there is a relationship between the variables.

Because significance levels indicate probabilities of chance occurrence, they are handy and intuitively appealing. To understand their logic, consider that a **probability** is a mathematical measurement of the likelihood that an event will occur in the future or has occurred in the past. As examples, one can assess the probability that a family with two children will have a boy and a girl, or the probability that a person in England in the 13th century died of the plague. Probabilities range from 0 to 1, where one means absolute certainty and zero indicates impossibility. An outcome with a probability of .5 means that it is as likely to happen as not. Probabilities are sometimes expressed in percentages. This simply requires multiplying the probability by 100.

Significance levels give the probability that chance produced an observed pattern in the data. Significance tests set thresholds for deciding whether the significance level is low enough to conclude that a relationship is real (not due to chance). If the significance level is low enough, the relationship is considered statistically significant. Different disciplines handle significance levels and statistical significance in different ways. Even the authors of this book vary in their approaches. Karen prefers to report significance levels because of their exacting detail. On the other hand, Steve prefers judging relationships as being present or absent if they meet the standard of .05, as is common in sociology. For beginning data analysts, it is useful to look at significance levels, but then judge them by the standards of significance tests.

At what threshold do significance levels become statistically significant? This is actually a complicated issue because it depends on the type of question a researcher is asking and the types of data the researcher is examining. To test relationships, norms of social research advocate conservative standards. Usually relationships are only considered real when chance is a very unlikely explanation of patterns in the data. Depending on the sample size, typically these thresholds are set at .05 and .01. These thresholds assert only a 5 in 100 chance (.05) or 1 in 100 chance (.01) that a pattern this strong would appear by chance. As you examine a relationship, look at the significance level. If it is below .10, conclude that the data *suggests* a real relationship, but a firm conclusion requires more research. A significance level of .05 or less, or .01 or less, more strongly supports the conclusion that a real relationship exists.

Before judging the statistical significance of any result, take a moment to think about significance levels. Seldom can the probability of an event be 0 (impossible) or 1 (inevitable). In a significance test, a significance level near 1.00 indicates that chance is almost certainly the

---

[13] To be precise, the significance level is the probability that a relationship appears in the sample, conditional on there being no relationship in the population. We offer the less precise, but more intuitive definition as few students (and some researchers) have trouble with the precise definition, but the two definitions offer the same conclusions.

cause of the relationship. In this situation, a researcher will have little confidence that the relationship between the variables is real. Conversely, a significance level close to 0, say .01, indicates that random chance is *extremely unlikely* to be the cause, and the researcher can be quite nearly certain that the relationship is real. Although chance is a possible cause, it is so unlikely that it is reasonable to rule out this explanation.

Interpreting Significance Levels

There are many methods for calculating significance levels, and we will not concern ourselves with mathematical formulas (feel free to breathe a sigh of relief!). Our interests are with interpretations. Interpreting significance levels requires understanding that they depend on two things: the **relationship strength** and **sample size**.

Relationship strength concerns how powerful associations are between two or more variables. Take, for instance the game of blackjack, a game in which the odds are stacked only slightly in favor of the house. If we watched professional blackjack players play the game for an hour, we would probably not see much of a relationship between the player, the house, and winnings. But if we watched roulette, a game in which the odds are stacked much more strongly in the house's favor, we would know who has the upper hand quite quickly, as the chips flow from the gamblers' hands into the casino's coffers at a much faster pace. Thus, the strength of the relationship (in this case who is favored in the rules of the game) influences how quickly the effects of random chance can be ruled out. But sample size – the number of observations - also has to be taken into account. We would not know who is favored in roulette on the basis of one or two spins of the wheel. And because the odds in blackjack are quite close between the house and the player, we would need many more observations in order to conclude with certainty that the rules operate in the house's favor. In sum, it is helpful to remember that small sample sizes require strong relationships to achieve significance. But if a sample is very large, even very small effects can be documented as reasonably beyond the influence of chance occurrence.

Let us pose another example. Suppose that Springfield's residents fear that their town's nuclear power plant is affecting their community's infant mortality rate. The residents ask a researcher to perform a study to determine if the nuclear power plant is at fault. One approach would be to compare Springfield's mortality rates over time. But if the researcher has only two samples, data from the year before the power plant was built, and the year afterward (a very small sample), the change in rates would have to be profound to attribute it to an association. In fact, it would be highly likely that the two rates would show small differences *even if the plant had no effect*, since mortality rates naturally fluctuate. Complicating the problem is that infant mortalities tend to be rare occurrences. However, if the rates had huge differences, the researcher would have greater confidence in the findings being real and not just an artifact of random fluctuation. Obtaining many years of data would strengthen the study considerably. A large enough sample size would render even a minute difference in infant mortality rates significant.

Significance levels and significance tests are vital in ruling out chance as the source of findings, but they have limits. When a relationship is statistically significant, it does not mean that the relationship is *important*. Even very small relationships can be statistically significant. Only knowledge of the variables involved can tell whether a relationship would actually affect anyone's life in a meaningful way. Furthermore, when a relationship is statistically significant, it does not mean that the relationship is *causal* (one variable causes the other). It may be that the direction of causality is reversed or that a third variable affected both variables simultaneously, creating a spurious relationship.

Analyzing Bivariate Relationships Between Two Categorical Variables

Performing bivariate analysis requires understanding the structure of variables, as described in Chapter 2. Recall that variables have two general forms—categorical or scale. This section addresses bivariate analysis between two categorical variables. Remember that categorical variables indicate typologies, such as "employment status," "marital status," or "sex." An individual is employed or unemployed; single, married, divorced, widowed or separated; and a man or a woman—these variables indicate into which category each individual falls. The methods described below show how to assess the relationship between one categorical variable and another categorical variable.

Cross Tabulations

Cross tabulations (abbreviated as **crosstabs**) show the association between two categorical variables in the form of a grid of all possible combinations of the two variables' categories. To create a cross tabulation, open the GSS04 data and run the following commands:

Analyze
Descriptive Statistics
Crosstabs
Rows: WRKSLF
Columns: SEX
OK

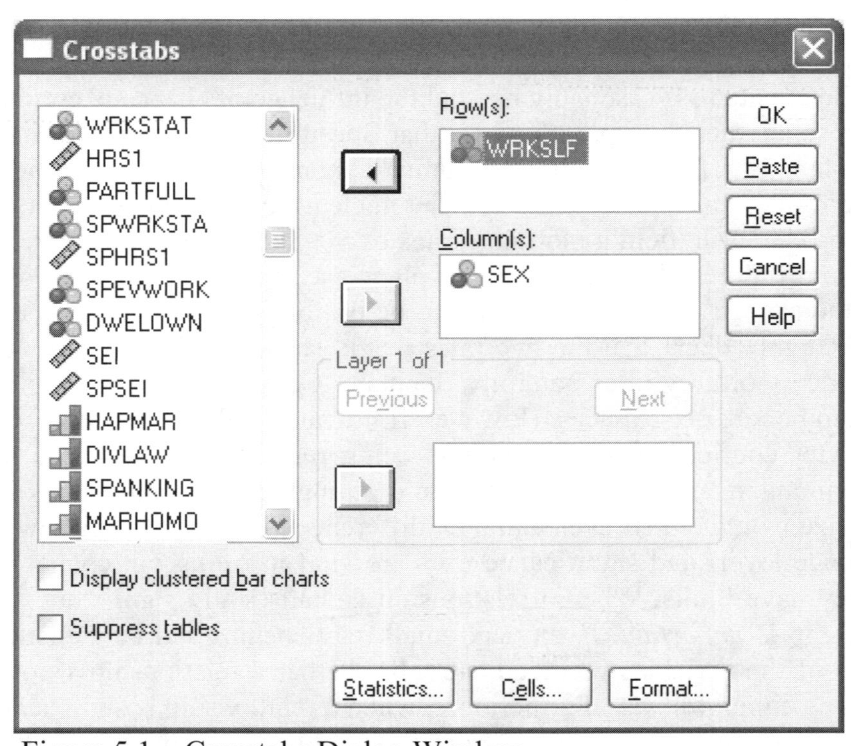

Figure 5.1 Crosstabs Dialog Window

The commands illustrated in Figure 5.1 create a table like the one in Figure 5.2, which reveals the number of men and women (in the two columns) and whether or not they are self-employed (in the two rows). This grid also shows how respondents are allotted in the "cells," the points of intersection between these two variables.

WRKSLF R SELF-EMP OR WORKS FOR SOMEBODY * SEX RESPONDENTS SEX Crosstabulation

Count

		SEX RESPONDENTS SEX		
		0 FEMALE	1 MALE	Total
WRKSLF R SELF-EMP OR WORKS FOR SOMEBODY	0 SOMEONE ELSE	1304	1049	2353
	1 SELF-EMPLOYED	141	202	343
Total		1445	1251	2696

Figure 5.2 A Crosstabulation Output With Cell Counts

Ultimately we want to use this crosstab to determine if men or women are more likely to be self-employed. This will require refining the output, but first let us consider the table in its representation of counts. This table splits the data into four groupings, with the respondents' gender in the columns, coupled with their employment statuses in the rows. Take a minute and look at this table. How many women in the sample are self employed? How many men work for someone else? You will also observe that the table gives totals, showing, for example, that there are a total of 2696 observations—1445 women and 1251 men. Pay attention to these counts in any analysis you run, as reliable analyses require sufficient observations in each cell of the table.

Now that you have studied the table, here is an analytic question. Who is more likely to be self-employed, men or women? You should be able to eyeball the table and conclude that men are more likely to be self-employed. But note that the study has more women than men, so it is not easy to compare the relative frequencies of self-employment between the genders. To do this, we will rerun the table and add a request that column percentages be included in the output.

Rerun the previous command, but specify that you want both the observed counts and column percentages reported. Your results should look like Figure 5.3.

Analyze
  Descriptive Statistics
    Crosstabs
      Rows: WRKSLF
      Columns: SEX
      Cells
        Counts: Observed
        Percentages: Column
        Continue
      OK

WRKSLF R SELF-EMP OR WORKS FOR SOMEBODY * SEX RESPONDENTS SEX Crosstabulation

			SEX RESPONDENTS SEX		
			0 FEMALE	1 MALE	Total
WRKSLF R SELF-EMP OR WORKS FOR SOMEBODY	0 SOMEONE ELSE	Count	1304	1049	2353
		% within SEX RESPONDENTS SEX	90.2%	83.9%	87.3%
	1 SELF-EMPLOYED	Count	141	202	343
		% within SEX RESPONDENTS SEX	9.8%	16.1%	12.7%
Total		Count	1445	1251	2696
		% within SEX RESPONDENTS SEX	100.0%	100.0%	100.0%

Figure 5.3 A Crosstabulation With Cell Counts and Column Percentages

Why did we ask for column percentages? The reason is that we want to compare the columns to each other: men to women. Think of it this way: imagine that we had all of the men and women in the study in an auditorium and wanted to create the same results by show of hands. To do this, we could corral all of the women on one side of the room, and all of the men on the other side. Then we would ask for the show of hands for who is self-employed and who works for someone else. The table works the same way.[14] It is easiest to construct tables consistently, so that the groups you are comparing are in the columns, and their characteristics are in the rows. At this point, you should have a much clearer interpretation of the data. Notice that 16.1% of men are self-employed, compared to only 9.8% of women.

But our work is not yet done. We need to see if this relationship is beyond the possibility of a chance occurrence in the data, explained by variation between the sample and the population it is expected to reflect. To do this, we will rerun the analysis yet again, but ask for one more statistic, the Chi-Square significance test.

Analyze
  Descriptive Statistics
    Crosstabs
      Rows: WRKSLF
      Columns: SEX
      Cells
        Counts: Observed
        Percentages: Column
        Continue
      Statistics
        Chi-square
        Continue
      OK

[14] Alternately, we could reframe the question to consider the percentages of self-employed people who are men and women. We could to this by replicating the above crosstabs, but then asking for row percentages instead of column percentages. Another way of doing this (the one we prefer) would be to switch WRKSLF to be the column variable and SEX to be the row variable, and then to request column percentages.

Crosstabs: Statistics

Chi-square
Correlations
Continue
Cancel
Help
Nominal
Contingency coefficient
Phi and Cramér's V
Lambda
Uncertainty coefficient
Ordinal
Gamma
Somers' d
Kendall's tau-b
Kendall's tau-c
Nominal by Interval
Eta
Kappa
Risk
McNemar
Cochran's and Mantel-Haenszel statistics
Test common odds ratio equals: 1

Figure 5.4 Crosstab Statistics Dialog Box

Chi-Square Tests

	Value	df	Asymp. Sig. (2-sided)	Exact Sig. (2-sided)	Exact Sig. (1-sided)
Pearson Chi-Square	24.651[b]	1	.000		
Continuity Correction[a]	24.079	1	.000		
Likelihood Ratio	24.624	1	.000		
Fisher's Exact Test				.000	.000
Linear-by-Linear Association	24.642	1	.000		
N of Valid Cases	2696				

a. Computed only for a 2x2 table

b. 0 cells (.0%) have expected count less than 5. The minimum expected count is 159. 16.

Figure 5.5 Chi-Square Output

Although SPSS displays the results for five tests, we limit our interest to the top one. This significance level is located in the column labeled "Asymp. Sig.," short for "Asymptotic Significance." Figure 5.5 shows that the significance level is .000, indicating a statistically significant relationship. There is less than a 1/1000 possibility that the observed relationship between sex and work situation is due to random chance. Although this value appears to be 0, a significance level of 0 is impossible. The value was actually rounded off to three digits after the decimal point. We can be very confident that there truly is a relationship between sex and self-employment status.

Now that we know there is a relationship, how can we measure its strength? Unfortunately, there is not a single, easily interpretable statistic that is appropriate for measuring the strength of all crosstabs. Some common statistics include Phi (for 2x2 tables in which both categorical variables are unordered), Cramer's V (for any sized table with unordered variables), and Gamma (for ordered categorical variables). Each of these statistics ranges from –1 to +1, with –1 indicating a strong negative relationship, 0 indicating no relationship, and +1 indicating a strong positive relationship. These statistics are available by clicking the "Statistics" button in the *Crosstabs* dialog box.

In this illustration of a crosstab, we broke the process into three component stages. First, we ran the crosstab and considered the cell counts and overall counts. There we eyeballed the data to see if all cells have reasonable numbers of observations and checked that there were sufficient observations to enable comparisons of the two categorical variables. Next, we considered the nature of the relationship by comparing percentages between the groups. Finally we addressed the issue of statistical significance, and were able to conclude whether the observed differences between men and women were the result of random chance, or if they reflected a real relationship. Of course, all the information is in the final output, which includes cell counts, percentages, and significance tests. Therefore, you can certainly dispense with the step-by-step creation of output, but not the consideration of all of the concerns outlined above.

Bar Charts

Bar charts display bivariate relationships between categorical variables. As an illustration, construct a bar chart showing the relationship between the variables in the previous cross tabulation. To do this, replicate the procedures below:

Graphs
Interactive
Bar

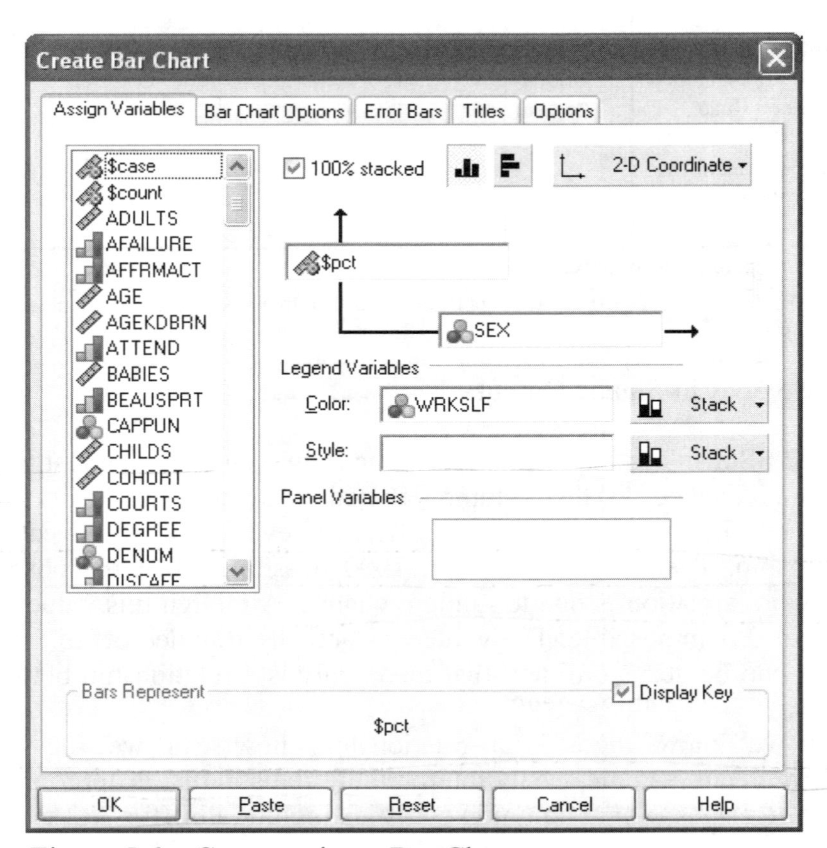

Figure 5.6 Constructing a Bar Chart

Note that in the dialog box, we asked SPSS to report percentages ($pct) and that to the right of the *Color* box into which we entered WRKSLF, we chose "*Stack*" by clicking on the small triangle. Making this specification required us to pull information from the "assign variables" panel, which was "drag and dropped" into the first dialog panel on the right. The variables WRKSLF and SEX can also be "drag and dropped." We also asked for the bars to be "*100% stacked*," so that it shows the proportions within men and women who are self employed. Had we not done this, the graph would scale bars according to their representation in the entire sample.

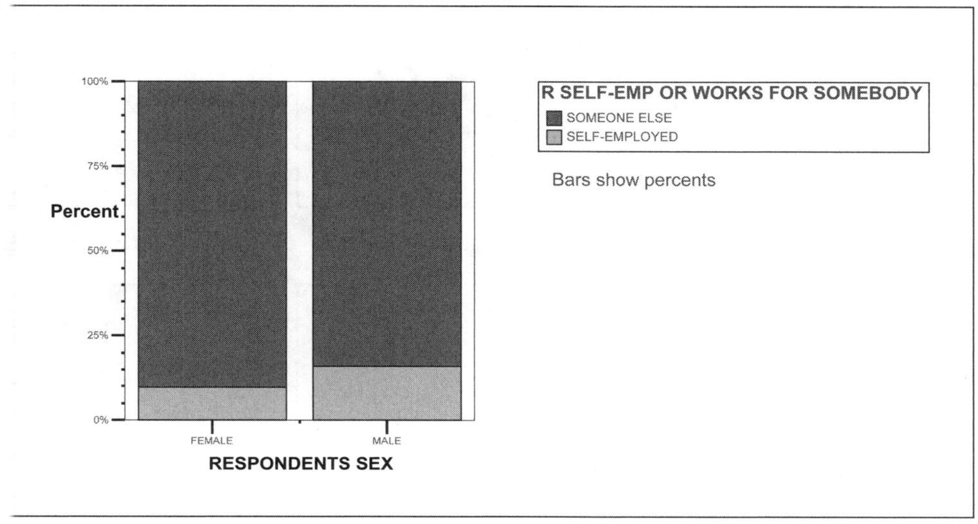

Figure 5.7 A Sample Stacked Bar Chart

Your bar chart will look slightly different than Figure 5.7. You can restructure your chart using the *Chart Editor* window by double clicking on the graph generated in the *Output Navigator* window, and editing the titles, bars, and other format concerns. The bar chart shows, first, that only a small percentage of men or women is self-employed, but also the degree to which men outnumber women in self-employment.

Analyzing Bivariate Relationships Between Two Scale Variables

Recall that scale variables indicate incremental amounts such as age, income, and miles driven to work. Because they fall on continua, these variables cannot be analyzed with the same approach as are categorical variables. For instance, if we tried to create a crosstab of the variables age and income, the crosstab could have literally thousands of cells, making meaningful comparisons impossible. But in some instances, scale variables can be recoded into categorical variables. In most situations, however, this is not the best approach, because the recoding procedure throws away much of the information contained within the scale. For example, by recoding income into categories, it becomes easier to compare rich and poor people, but the capacity to compare differences *within* the rich and poor categories is washed away. Therefore, when analyzing scale variables, keep them in their original format, and only create categories when there is clear theoretical or analytic justification for doing so. In this section, we consider how to assess the relationships between two scale variables, situations in which both variables have observations arrayed on continua.

Correlations

Most researchers use **correlations** to summarize the association between two scale variables. Generally, when social scientists discuss correlations, they are referring to **Pearson's correlation coefficient**. Pearson's correlation coefficient (from here on the "correlation") measures the strength of the **linear relationship** between two variables.

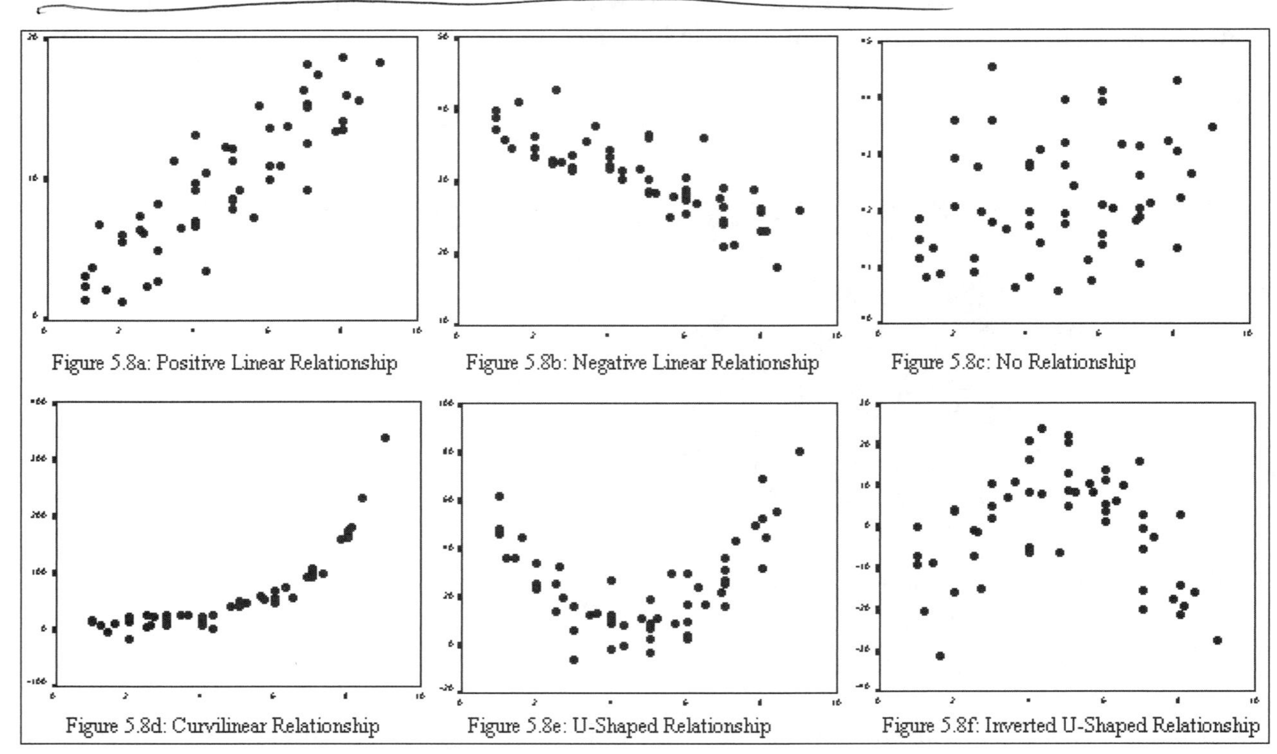

Figure 5.8a: Positive Linear Relationship
Figure 5.8b: Negative Linear Relationship
Figure 5.8c: No Relationship
Figure 5.8d: Curvilinear Relationship
Figure 5.8e: U-Shaped Relationship
Figure 5.8f: Inverted U-Shaped Relationship

Figure 5.8 Various Types of Relationships Between Scale Variables

Linear relationships are only one kind of relationship between scale variables. Other possibilities include **curvilinear**, as well as **"U" shaped** and inverted "U" shaped relationships (see Figure 5.8). Correlations identify linear relationships, but researchers should always be sensitive to the possibilities of other kinds of relationships. For example, the most likely relationship between test anxiety and test performance is an inverted U. People whose anxiety is very low (mentally sleepy) or very high (overwrought with anxiety) perform worse than those whose anxiety is moderate (alert and sharp).

Linear relationships exist when a difference in the value of one variable is associated with a consistent difference in the value of another variable. For example, if each additional year spent in school is associated with a $1,000 increase in starting salary following graduation, the relationship between education and income would be linear. If the salary increase were $1,600 for each additional year of schooling at the low end of the scale, but only $900 for each additional year of schooling at the high end, the relationship would not be linear.

The correlation coefficient can have values from -1 to +1. A correlation of 0 indicates that there is no linear relationship between the two variables. Correlations of -1 and +1 indicate that there is a perfect linear relationship between the two variables. The closer the correlation is to either +1 or -1, the stronger the relationship between the two variables. Conversely, the closer the correlation is to 0, the weaker the correlation. In the social sciences, a correlation of .30 using individual level data is considered a "good" correlation; a correlation above .40 is considered "strong." Group level data, as in the STATES07 data set, have higher thresholds (.50 might be considered "good" and .65 or more "strong.") Some examples of linear relationships and their correlations are displayed in Figure 5.9.

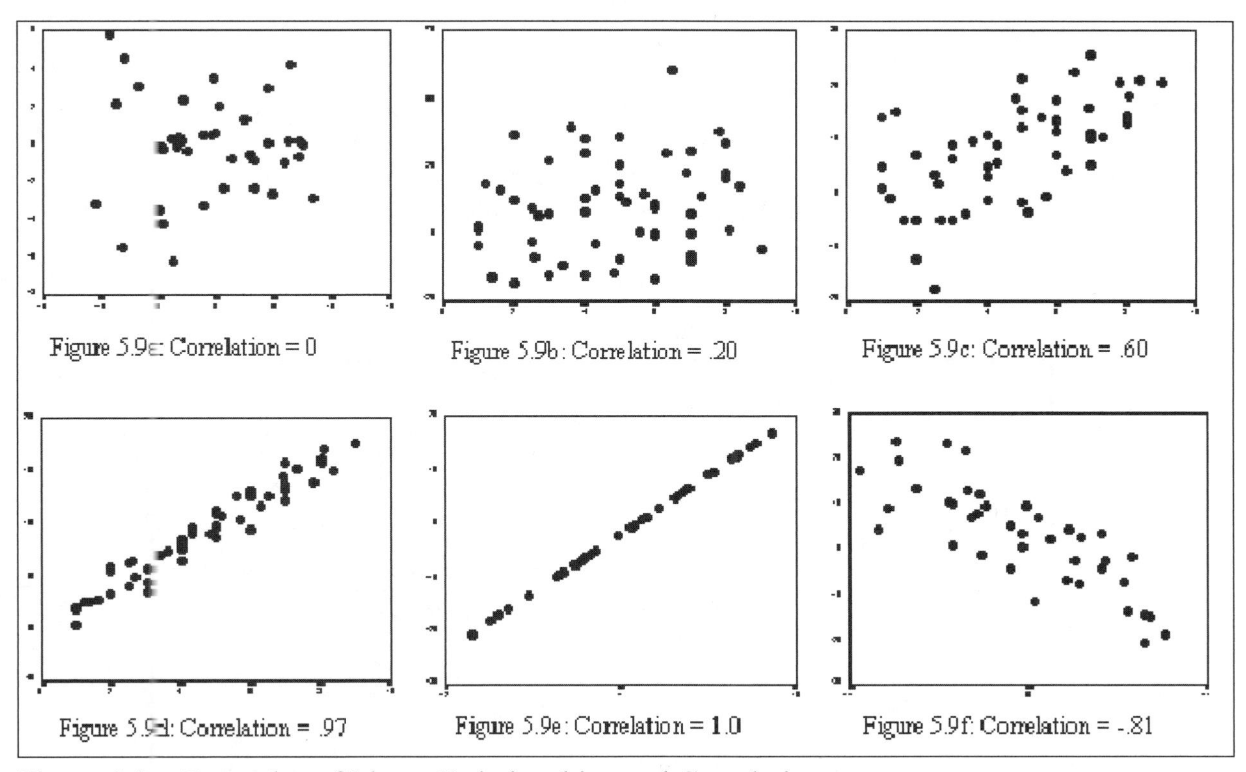

Figure 5.9 Examples of Linear Relationships and Correlations

The negative and positive signs indicate the direction of the relationship. A positive correlation means higher values of one variable are associated with higher values of the other variable. There is a positive correlation between education and income—more educated people tend to have higher incomes. A negative correlation, on the other hand, means higher values in one variable are associated with lower values in the other variable. There is a negative correlation between education and body weight—more educated people have generally lower body weights.

To illustrate the application of correlation, we will examine the relationship between "social disorganization" and suicide. This relationship was identified by sociologist Emile Durkheim (1897), who found that social forces that exist external to individuals also influence them. He suggested that places with high levels of social disorganization (e.g., crime, divorce, substance abuse) also have high levels of suicide. Social disorganization is a construct that is measured by a variety of indicators that reveal the strengths of social ties between individuals and the society. One such measure is the divorce rate. According to Durkheim's theory, people who reside in areas with high divorce rates live in environments with greater social disorganization than those in areas with low divorce rates. This is not to say that divorce causes suicide, or that people who get divorced are the ones committing suicide, only that loose social bonds (as indicated by high levels of divorce) are theoretically associated with extreme behaviors such as suicide.

We can test Durkheim's hypothesis using the STATES07 data set. Open this data set and run a correlation on the relationship between suicide rates and the divorce rate. If Durkheim is correct, we should expect to find a positive correlation between these two variables.

Analyze
Correlate
Bivariate:
Variables: HTH179 (Death Rate by Suicide 2003)
DMS483 (Divorce Rate 2005)
Correlation Coefficients: Pearson
Tests of Significance: Two Tailed
OK

Correlations

		HTH179 Death Rate per 100,000 by Suicide: 2003	DMS483 Divorce Rate per 1,000 Pop.: 2005
HTH179 Death Rate per 100,000 by Suicide: 2003	Pearson Correlation	1	.720**
	Sig. (2-tailed)		.000
	N	51	46
DMS483 Divorce Rate per 1,000 Pop.: 2005	Pearson Correlation	.720**	1
	Sig. (2-tailed)	.000	
	N	46	46

**. Correlation is significant at the 0.01 level (2-tailed).

Figure 5.10 Correlation Output

Figure 5.10 shows correlation output, which displays correlations among every variable specified. The two variables we specified, DMS483 and HTH179, make up both the columns and the rows. There are four boxes because each box contains information about the correlation between the column variable with the row variable. Every correlation table contains correlations of 1 along the diagonal from top left to bottom right, since these boxes represent each variable's correlation with itself. Furthermore, the table is symmetric above and below this diagonal—the same information is displayed. In this small correlation table, the upper right and the lower left box are the same, because both boxes contain the correlation between HTH179 and DMS483. Out of these four boxes, we only need to look at the upper right.

In each box the top number is the correlation coefficient, which represents the strength of the relationship. Recall that this number will always be in a range of -1 to +1. The further that number is from 0, the stronger the relationship. The second number in each box is the significance. Recall that the closer this number is to 0, the lower the likelihood that the relationship can be attributed to the effects of chance. The third number is the sample size (N). It is prudent to check that the sample does not have too many excluded cases. Here we see that information on this relationship exists for 46 out of the 51 states (including Washington DC).

The correlation between divorce and suicide offers strong support of Durkheim's thesis. There is a .72 correlation between these two variables, indicating a positive relationship between suicide and divorce rates. This relationship is statistically significant as well. This can be seen in two ways from the Correlations table. The significance test shows a probability of .000, indicating that this is a statistically significant relationship. Also, the correlation tables flag out significance with asterisks (**) next to the coefficients.

Again, it is important to emphasize what correlations can and cannot do. Correlations show how much two numerical variables "co-relate" in a linear fashion. They cannot measure other types of relationships, such as curvilinear relationships. The U-shaped relationships in Figure 5.8 would have correlations of 0. Further, a correlation does not mean that one variable causes the change in the other variable to occur, only that their values are associated. Finally, correlations in the STATES07 data reflect patterns revealed in places, not individuals. To assume that these group patterns are revealed in the lives of individuals is to engage in the **ecological fallacy**, which errantly asserts an individual's motivations and experiences from group measures. In the above example, an ecological fallacy would be to conclude that those experiencing divorces are more likely to kill themselves. We only know that places with higher levels of divorce also tend to be places with higher levels of suicide.

Scatterplots

Scatterplots create visual images of relationships between two scale variables, and are a useful complement to correlations. The graphs in Figures 5.8 and 5.9 are examples of scatterplots. In these graphs, the independent variable is usually placed on the X axis (horizontal axis) and the dependent variable is placed on the Y axis (vertical axis). Each case is placed on the graph at the intersection of its values for the two variables. Because we are testing the effects of divorce rates on suicide rates, we plot divorce rates (the independent variable) on the X axis and suicide rates (the dependent variable) on the Y axis.

To create a scatterplot, replicate the dialog box illustrated in Figure 5.11

Graphs

Interactive

Scatterplot

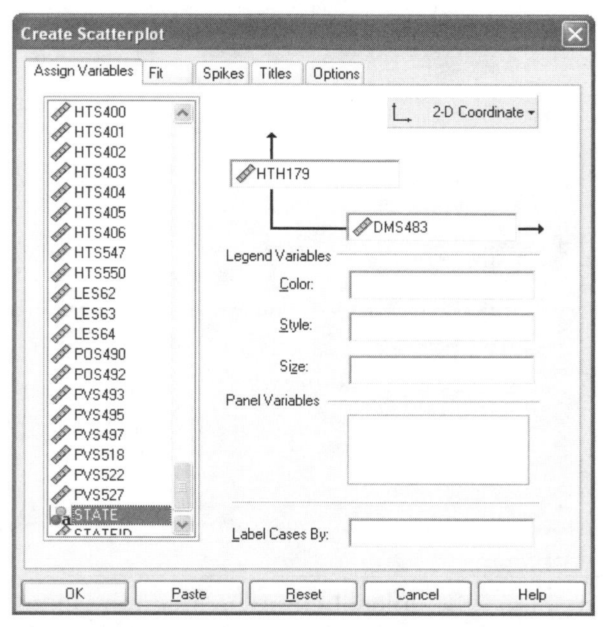

Figure 5.11 A Scatterplot Dialog Window

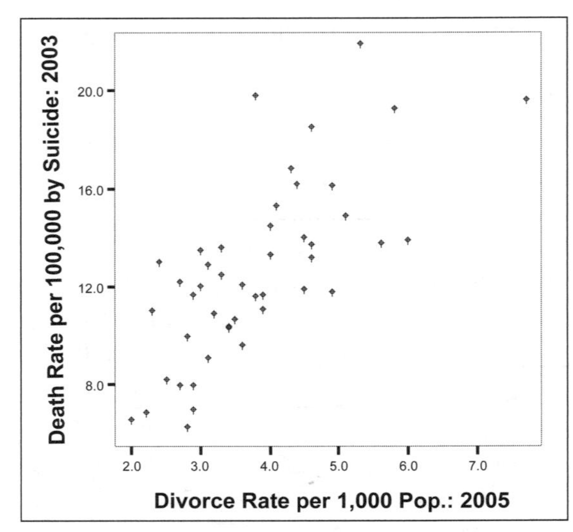

Figure 5.12 Scatterplot output

Finally, we offer a few ways to modify and refine scatterplots. One is to change the symbols within the scatterplot. To do this, "double click" on the graph to open an editing window and then "right click" on any observation within the graph. This will enable you to change the appearance of the symbols within the graph with the options on the left side of the chart editor window. This is illustrated in Figure 5.13. You can also specify that a specific variable within the data set to be used to mark observations. In this case, we might want to label each observation by its state name. This process is illustrated in Figure 5.14 and the output is reproduced in Figure 5.15.

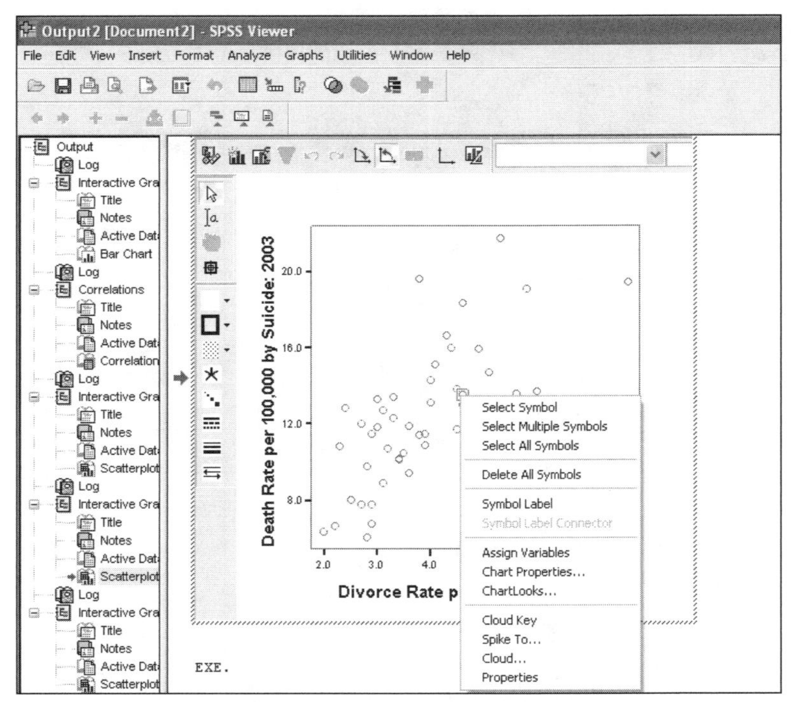

Figure 5.13 Double Clicking on a Graph Opens a Chart Editor Window. Right Clicking on a Symbol Opens Options for Displays

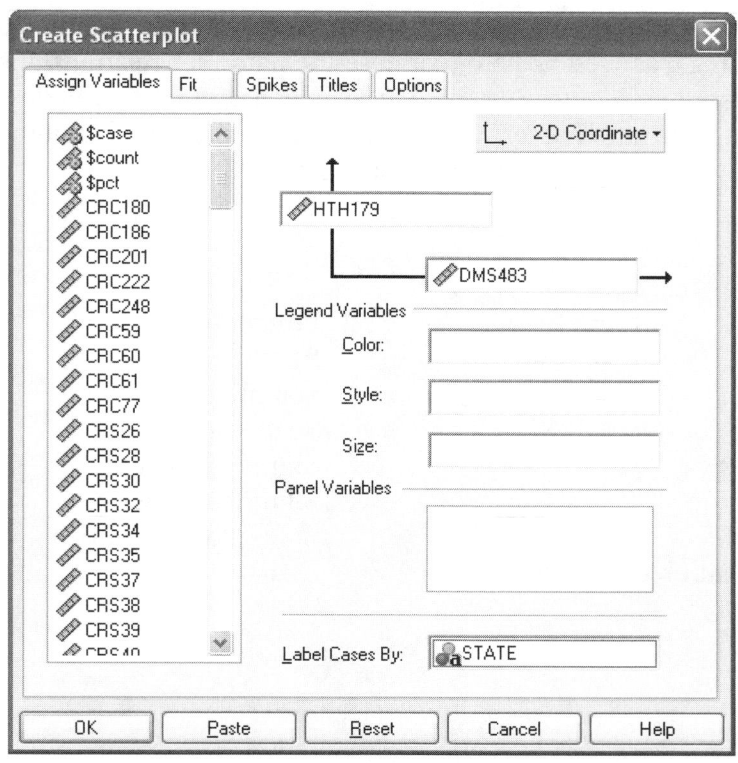

Figure 5.14 Scatterplot Dialog Window with a Label Cases Command

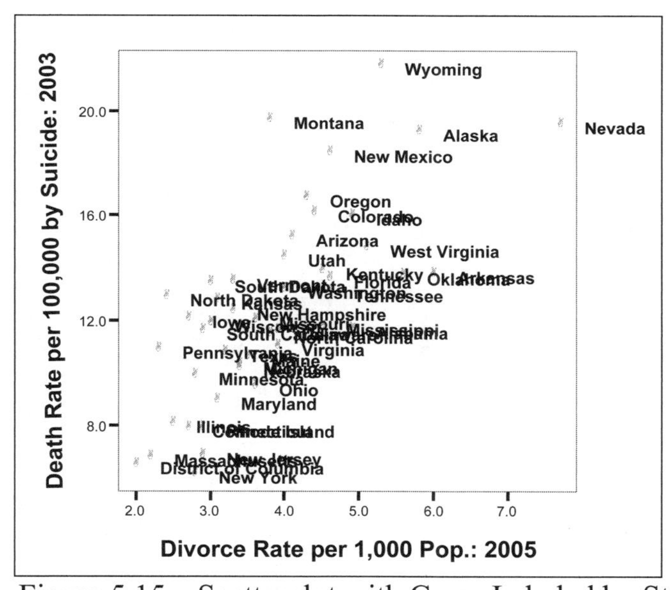

Figure 5.15 Scatterplot with Cases Labeled by State Names

Summary

This chapter demonstrated methods of verifying the existence of statistical associations using SPSS through crosstabs and correlations. Each method has accompanying significance tests that inform the researcher if associations are strong enough that chance can be ruled out as a likely explanation for observed relationships.

In addition, we showed how to graph bivariate relationships and create visual depictions that can present findings, as well as help diagnose the nature of relationships. Bivariate analyses are important because they identify the degree to which any two variables are associated with one another. In the following chapter, we extend this discussion, examining ways of analyzing the relationships between a categorical and a scale variable.

Key Terms

Alternative hypothesis
Bar charts
Correlations
Cross tabulations
Curvilinear relationship
Ecological fallacy
Linear relationship
Null hypothesis
Pearson's correlation coefficient
Probability
Random sample
Sample size
Scatterplots
Significance level
Significance tests
Statistically significant
U shaped relationship

Chapter 5 Exercises

Name____________________________________ Date______________________

1. A police force has increased its employment from five full-time officers in 2006 to eight full-time officers in 2007 in an effort to deter crime. They call you in to analyze whether their program has been effective and give you the following output to analyze. How would you interpret this cross tabulation to the police department?

	Number of Arrests	
	2006	2007
Property Crime Arrests	325	375
Violent Crime Arrests	120	119

Sig. = .03

2. A researcher examines a relationship between "sensitivity" and "cultural acceptance." As a result of her study she finds a correlation of .20 between these variables and a significance level of .25. On the basis of this study, can she conclude that sensitivity is associated with cultural acceptance? Explain.

3. Using the GSS04 data, examine the relationship between attitudes toward the level of national assistance for childcare (NATCHLD) and the sex (SEX) of the respondent. Make a bar chart of the relationship. Fill in the following information:

   Percentage of men stating the current level is too little ________________

   Percentage of women stating the current level is too little ________________

   Chi square significance level ________________

   Is the relationship statistically significant? Yes No

   How would you interpret this result?

4. Using the GSS04 data, examine the relationship between race (RACE) and whether a person supports the death penalty for murder (CAPPUN). Make a bar chart of the relationship.

Percentage of Whites opposed ____________________

Percentage of African Americans opposed ____________________

Percentage of others opposed ____________________

Chi square significance level ____________________

Is the relationship statistically significant? Yes No

How would you interpret this result?

5. Using the GSS04 data, examine the relationship between race (RACE) and the belief that Whites are hurt by affirmative action (DISCAFF). Make a bar chart of the relationship.

Percentage of Whites saying
"Somewhat Likely" or "Very Likely" ____________________

Percentage of Blacks saying
"Somewhat Likely" or "Very Likely" ____________________

Percentage of others saying
"Somewhat Likely" or "Very Likely" ____________________

Chi square significance level ____________________

Is the relationship statistically significant? Yes No

How would interpret this result?

6. Using the GSS04 data, create a crosstab examining the relationship between HAPPY (general happiness) and MARITAL (marital status). Do you observe a relationship between people's marital status and their level of happiness? Test this relationship with the appropriate significance test. Can you conclude that there is a relationship?

Percentage of married individuals are "very happy" ____________

Percentage of divorced individuals who are "very happy" ____________

Percentage of never married individuals who are "very happy" ____________

Chi square significance level ____________

Is the relationship statistically significant? Yes No

How would interpret this result?

7. Is there a relationship between a person's age (AGE) and the number of hours spent per day watching TV (TVHOURS)? Use the GSS04 data to test this relationship.

Correlation	____________	
Significance level	____________	
Is the relationship statistically significant?	Yes No	
Do people tend to watch more or less TV as they age?	More	Less

Make and print a scatterplot of the relationship.

How would you explain these findings?

8. Using the STATES07 data, perform a test to see if there is a relationship between the percent of the population that graduated from high school (EDS129) and the median earnings of male full-time workers (EMS161) in the state.

   Summarize and interpret your findings below.

   Correlation ____________

   Significance level ____________

   Is the relationship statistically significant? Yes No

   Make and print a scatterplot of the relationship.

   How does the percent graduated relate to the median earnings of female full-time workers (EMS162)? Is the relationship the same as for male workers?

   How would you explain these findings?

9. Is there a relationship between the average salary of classroom teachers (EDS121) and the percent of population that has graduated from high school (EDS129)? Use the STATES data to examine this question. Summarize and interpret your findings below.

Correlation ____________

Significance level ____________

Is the relationship statistically significant? Yes No

Make and print a scatterplot of the relationship.

In what ways are teacher salaries associated with high school graduation rates?

How would you explain these findings?

10. Pose a hypothesis between any two scale variables in either the STATES or GSS04 data sets.

A. State the hypothesis:

B. Construct and print a refined univariate graph of the dependent variable.

C. Construct and print a refined bivariate graph of the relationship between these two variables.

D. Determine if the relationship is statistically significant. Sig.=_____________

Is it significant? Yes No

E. Explain the extent to which your hypothesis is supported by your analysis.

Chapter 6
Comparing Group Means through Bivariate Analysis

Overview

In the previous chapter we considered strategies to relate scale variables with each other, and categorical variables with each other. This chapter extends this discussion by considering methods that relate scale variables to categorical variables. New statistical techniques are required to combine information from variables with different structures. These include comparing group means using of analysis of variance, t tests, post-hoc tests, and visually comparing group characteristics through bar charts and box plots.

One-Way Analysis of Variance

Answering some of the most interesting questions in social science research requires comparing groups, or classifications, of observations on the basis of incremental measures. Consider life expectancies. According to the Statistical Abstracts of the United States, as a group, women outlive men by 5.2 years, and White Americans outlive African Americans by 5 years.[15] These comparisons required dividing the population into categories (men/women or White/Black), and then considering how much the categories vary on a scale variable (in this case years of life expectancy).

Gender, race, educational attainment, geographic location, marital status, and a variety of other designations divide the world into categorical groups. And these groups relate with a variety of scale variables, not only in terms of life expectancy, but also income, weight, height, alcohol consumption, to name a few examples. The methods we address in this chapter concern comparing values of scale variables between categorical groups.

[15] In 2004, the life expectancy at birth in 2004 was 75.2 years for males, 80.4 years for females, 78.3 years for Whites, and 73.3 years for African Americans. African American men had a life expectancy of only 69.8 years.

One of the most common ways to assess differences between groups is to perform **one-way analysis of variance (ANOVA)** tests. These compare the mean values of a scale variable as they differ among the categories of a categorical variable, as well as test these relationships for statistical significance.[16] To illustrate the application of one-way ANOVA, we will examine if television viewing differs by educational attainment. Our hypothesis is that educational attainment is negatively related to television viewing. In other words, more educated people tend to spend less time watching television, and the group that watches the most television are those who lack a high school degree. To test this hypothesis, we will use the GSS04 data set using TVHOURS, a scale variable measuring the daily number of hours spent watching television and DEGREE, an ordinal categorical variable measuring whether each person attained less than high school, a high school degree, a junior college degree, a bachelors, or a graduate degree. Notice that in this example, it would not be appropriate to use correlations or cross tabulations, although recoding TVHOURS into categories (e.g., none, 1-3, 4-6, more than 6 hours) could allow a cross tabulation. Our goal is to find out the mean number of hours members of each group watch television, and to test whether any observed differences between groups are statistically significant (reasonably beyond the bounds of chance).

To test the relationship between television viewing and educational attainment with ANOVA, use the following commands:

Analyze
Compare Means
One-way ANOVA
Dependent List: TVHOURS
Factor: DEGREE
Options
Statistics: Descriptive
Continue
OK

We concentrate on the Descriptives table (Figure 6.1) and the ANOVA table (Figure 6.2) of the ANOVA output, both of which address the research question.

Descriptives

TVHOURS HOURS PER DAY WATCHING TV

	N	Mean	Std. Deviation	Std. Error	95% Confidence Interval for Mean		Minimum	Maximum
					Lower Bound	Upper Bound		
0 LT HIGH SCHOOL	111	4.50	3.697	.351	3.80	5.19	0	20
1 HIGH SCHOOL	476	2.96	2.561	.117	2.73	3.19	0	20
2 JUNIOR COLLEGE	79	2.53	1.894	.213	2.11	2.96	0	12
3 BACHELOR	141	2.17	2.080	.175	1.82	2.52	0	15
4 GRADUATE	92	1.78	1.333	.139	1.51	2.06	0	6
Total	899	2.87	2.617	.087	2.69	3.04	0	20

Figure 6.1 Descriptives Table in ANOVA

[16] There are other types of ANOVA, including two-way ANOVA, which are beyond the scope of this book.

ANOVA

TVHOURS HOURS PER DAY WATCHING TV

	Sum of Squares	df	Mean Square	F	Sig.
Between Groups	483.655	4	120.914	19.075	.000
Within Groups	5667.059	894	6.339		
Total	6150.714	898			

Figure 6.2 ANOVA Table Revealing Statistical Significance

First we look at the *Descriptives* table. The third column, labeled "Mean," shows that people with less than a high school education spend 4.50 hours watching television each day, and that the means are smaller for each higher degree. As hypothesized, those with graduate degrees spend the least time watching television (1.78 hours a day). But our work is not yet done. We will need to check whether the differences among the means are significant in the ANOVA table, and also examine the rest of the *Descriptives* table to make sure everything is as it should be.

The second column, labeled "N" shows the sample size for each group. Just as in crosstabs, the intersection of two variables has the potential to create small groups. Of the 899 people who report both degree and television viewing, the smallest group is 79 people with a junior college degree, a reasonably large sample size. We conclude that the analysis is not threatened by small sample sizes. Were there only a few people in a category, we could collapse categories together or exclude categories with low representation from the analysis.

The columns for standard deviation, standard error, and confidence intervals all describe the spread of values in each group. As we discuss in more detail below, one assumption of ANOVA is that the categories have reasonably equal spreads. The standard deviations are all quite similar—none is so different from the others as to raise any doubts at this point about the integrity of the analysis.

Finally we examine the significance tests in Figure 6.2 to determine if the relationship should be attributed to a real statistical relationship, or if it could simply reflect a chance association, which would likely disappear if we had the opportunity to replicate this study with a new sample. The ANOVA table reports the significance level (Sig.) is .000. This statistic can be interpreted in the same way we interpreted other significance statistics — it is highly unlikely that the differences in means occurred by chance. We therefore conclude that there is a relationship between educational attainment and television viewing. Of course we take caution not to assert that this is a causal relationship, something that ANOVA (alone) cannot determine.

Post-Hoc Tests

Often it is important to not only determine *if* differences exist among the categories, but *where* those differences exist. In this section we consider how to test for differences in the television viewing habits among the different degree groups. In the findings above, we determined that an association exists between television viewing and educational attainment. But notice that some of the groups have mean scores that are quite similar to one another. For example, people with junior college degrees (2.53 hours) appear to spend more time watching television than those with bachelors degrees (2.17 hours). But is this difference large enough for us to conclude that the television viewing habits of these two groups are significantly different from one another? The answer requires a post-hoc test. **Post-hoc tests** indicate *which* group means differ from one another. This is an important refinement to the analysis, as it goes beyond the basic conclusion that a difference exists somewhere among all of the groups. To add a post-hoc test to the ANOVA, perform the following command:

Analyze
Compare Means
One-way ANOVA
Dependent List: TVHOURS
Factor: DEGREE
Options: *Statistics*: Descriptive
Continue
Post-hoc: *Equal Variances Assumed*: Tukey
Continue
OK

Multiple Comparisons

Dependent Variable: TVHOURS HOURS PER DAY WATCHING TV
Tukey HSD

(I) DEGREE RS HIGHEST DEGREE	(J) DEGREE RS HIGHEST DEGREE	Mean Difference (I-J)	Std. Error	Sig.	95% Confidence Interval	
					Lower Bound	Upper Bound
0 LT HIGH SCHOOL	1 HIGH SCHOOL	1.540*	.265	.000	.81	2.26
	2 JUNIOR COLLEGE	1.964*	.371	.000	.95	2.98
	3 BACHELOR	2.325*	.319	.000	1.45	3.20
	4 GRADUATE	2.713*	.355	.000	1.74	3.68
1 HIGH SCHOOL	0 LT HIGH SCHOOL	-1.540*	.265	.000	-2.26	-.81
	2 JUNIOR COLLEGE	.424	.306	.636	-.41	1.26
	3 BACHELOR	.786*	.241	.010	.13	1.45
	4 GRADUATE	1.173*	.287	.000	.39	1.96
2 JUNIOR COLLEGE	0 LT HIGH SCHOOL	-1.964*	.371	.000	-2.98	-.95
	1 HIGH SCHOOL	-.424	.306	.636	-1.26	.41
	3 BACHELOR	.361	.354	.845	-.61	1.33
	4 GRADUATE	.749	.386	.297	-.31	1.80
3 BACHELOR	0 LT HIGH SCHOOL	-2.325*	.319	.000	-3.20	-1.45
	1 HIGH SCHOOL	-.786*	.241	.010	-1.45	-.13
	2 JUNIOR COLLEGE	-.361	.354	.845	-1.33	.61
	4 GRADUATE	.388	.337	.780	-.53	1.31
4 GRADUATE	0 LT HIGH SCHOOL	-2.713*	.355	.000	-3.68	-1.74
	1 HIGH SCHOOL	-1.173*	.287	.000	-1.96	-.39
	2 JUNIOR COLLEGE	-.749	.386	.297	-1.80	.31
	3 BACHELOR	-.388	.337	.780	-1.31	.53

*. The mean difference is significant at the .05 level.

Figure 6.3 Multiple Comparisons Output

You will notice that the Post-hoc window in ANOVA makes it clear that there are many post-hoc tests to choose from. Although their methods are different, they generally test the same thing. A **Tukey test** is one of the most commonly used post-hoc tests, and is the one we illustrate. The results are replicated in the Multiple Comparisons table in Figure 6.3. This output provides a comparison of the differences in the average (mean) viewing hours of all groups. The table is large, but we will focus on the "mean difference" column, which gives the difference in means for each pair of categories that are being compared. Two groups with equal means would have a mean difference of 0. Positive mean difference indicates that the first group listed (in column I) has a higher mean; a negative mean difference means that the second group (in column J) has a higher mean. Significant mean differences (with significance levels below .05) are marked with a "*". The exact significance coefficient is in the "Sig." column.

Consider the television viewing habits of respondents with less than a high school education (the first row of statistics). The first mean difference shows that respondents with less than a high school education watch, on average, 1.54 hours more television per day than respondents with a high school degree. They also spend 1.964 more hours watching television than junior college graduates, 2.325 hours more than bachelor's degree recipients, and 2.713 hours more than graduate degree holders. All of these differences are statistically significant at a level of .000, supporting our hypothesis that people with less than a high school degree watch more television than others.

But now note something interesting in the other rows of statistics - the lack of significant differences between those with junior college, bachelor's degrees, and graduate degrees. This finding suggests that having any college degree is associated with lower amounts of television viewing, but that additional education from the associate degree upwards has no significant effect on television viewing behavior. While the mean difference scores show a tendency in the expected direction, this is not a strong enough association to discount chance as a plausible explanation.

Assumptions of ANOVA

As we discuss below, and in the remaining chapters in this book, advanced statistical procedures, such as ANOVA, operate under specific assumptions about the origin and structures of the categorical and scale variables. Although ANOVAs and post-hoc tests are useful, their results are only trustworthy if the variables satisfy certain conditions. It is the researcher's responsibility to make sure that the data meet the assumptions of the test. ANOVA assumes the samples are independent, and each group's distribution is normal with the same standard deviation, concerns we address below.

Independence Assumption

One assumption of one-way ANOVA is that the observations in different categories are **independent** of each other. This means that the observations in one category are not paired up in some way with the observations in any other category. For example, if we analyzed television viewing by gender, and if both husbands and wives were interviewed in the study, measures of television viewing for married couples would be interdependent. Because husbands and wives form television-viewing habits together, they cannot comprise independent observations. However, ANOVA will treat them as such. In the case of the GSS04, it is unlikely that any observations are interdependent, as the study relies on a random sample of individuals who are unlikely to be connected with each other in meaningful ways.

Normality Assumption

In ANOVA the observations of the scale variable within each category should have a normal, or bell-shaped distribution. In practice, distributions that are slightly skewed generally do not pose a problem for ANOVA. However, distributions that are highly skewed or U-shaped do pose a problem, and ANOVA is not the appropriate way to analyze such cases. Because the application of advanced data analytic techniques relies on understanding the structure of variables, this highlights the need to implement the techniques learned earlier in univariate analysis. In other words, performing successful bivariate analyses such as ANOVA requires foundational work to understand variables in isolation from one another. For instance, forming a histogram of television viewing is one way of checking that this variable reasonably conforms to the normality assumption. Recall that performing the following can test this assumption:

Graphs
Interactive
Histogram
Assign Variables:
$COUNT
TVHOURS
Histogram
Normal Curve
OK

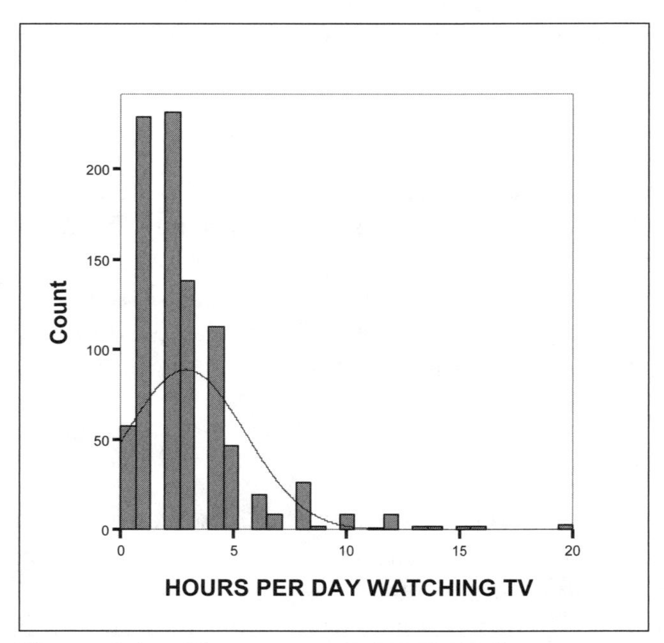

Figure 6.4 Histogram of TVHOURS

As Figure 6.4 shows, television viewing has a skew. One would want to keep this in mind when applying advanced techniques that operate on normality assumptions. In practice, the distributions of many variables appear like that of TVHOURS. In this case, our main concern would be the small number of cases that are watching 15 or more hours of television per day. We would want to make sure that they do not exert too much leverage on mean-based analyses such as ANOVA. One means of checking this is to run two sets of ANOVAs, one with these cases included in the analysis, and another set of runs with them excluded.

Graphing the Results of an ANOVA

Two ways of graphing group comparisons are bar charts and box plots. Bar charts are familiar to many people and are appropriate for presenting findings to the general public or those not well versed in statistical analysis. Box plots, however, offer more detailed information, showing both the central tendencies and distributions of each category.

Bar Charts

Bar charts allow a researcher to quickly compare the means of different categories. The categories are placed along the horizontal axis and the height of each bar is the mean of the scale variable within that category. To construct a bar chart showing how television viewing varies by degree, use the following commands:

Graphs
Interactive
Bar
Assign Variables
TVHOURS
DEGREE
Bars Represent TVHOURS *Means*
OK

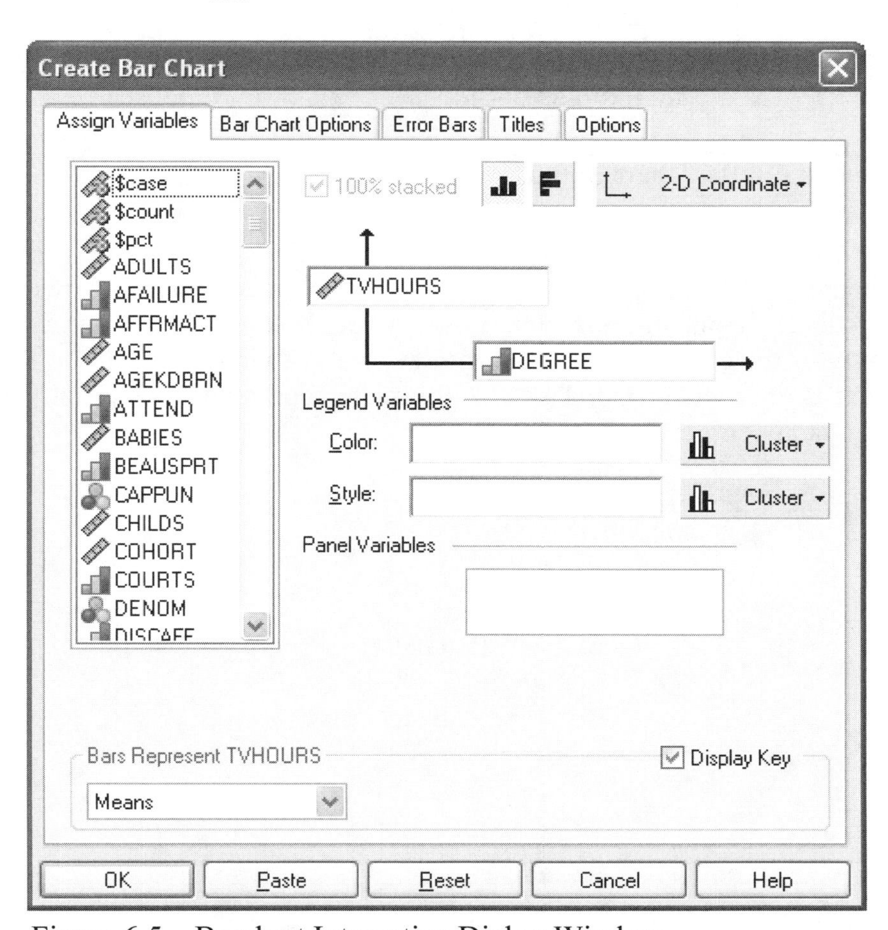

Figure 6.5 Barchart Interactive Dialog Window

The resulting bar chart, in Figure 6.6, clearly shows that people with less than a high school education watch more television, on average, than all other groups. Notice that the bars replicate the findings reported in the ANOVA table. For example, people with less than a high school education watch, on average, 4.5 hours of television per day.

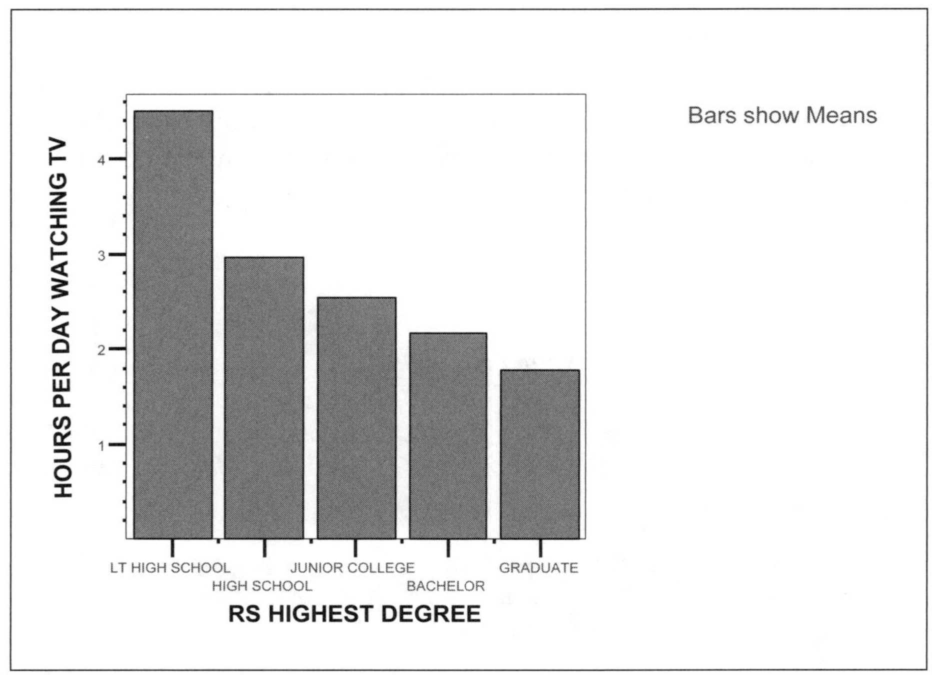

Figure 6.6 Bar Chart Output

Box Plots

Recall from Chapter 3 that a box plot displays a variable's distribution, showing the spread of cases and the median. By placing box plots side-by-side, it is possible to get a very good visual depiction not only of the central tendencies of television viewing, but also the range of viewing behavior that exists in each group. Use the following commands to construct box plots of television viewing and degree:

Graphs
Interactive
Boxplot
TVHOURS
DEGREE
OK

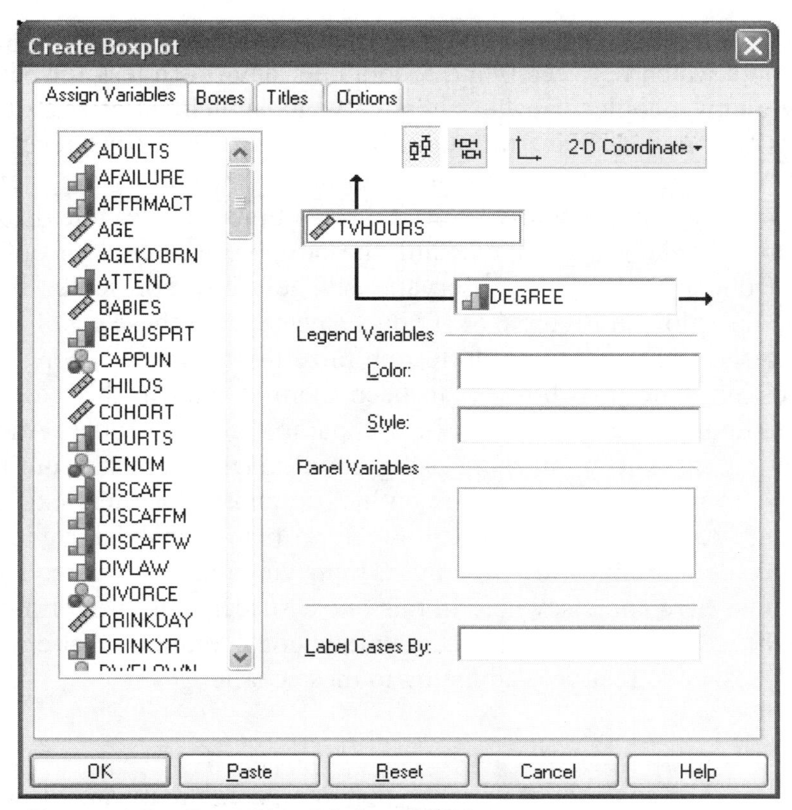

Figure 6.7 Box Plot Dialog Window

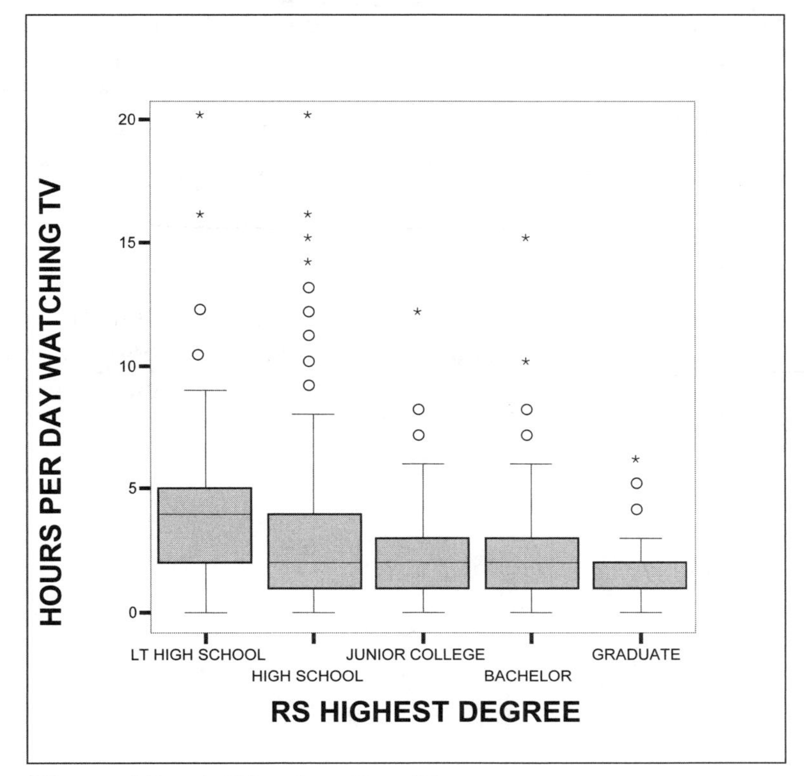

Figure 6.8 A Bivariate Box Plot

Figure 6.8 illustrates a bivariate box plot of TVHOURS and DEGREE. These box plots show that most people watch between 0 and 8 hours per day (even less for people with graduate degrees), but that a few people in each category watch much more television. One thing to remember is that a box plot displays the median, not the mean, so its results will only match those of the ANOVA if the distributions are symmetric. Box plots are therefore particularly informative when variables have slightly skewed distributions, as does television viewing.

Box plots are also helpful in diagnosing the possible influence of outliers, observations that are polarized from the rest of the observations. These are represented by the small circles and asterisks in the box plots in Figure 6.8. If you suspected that these cases are having a strange effect on your analysis, what could you do to minimize their impact? There is no single correct answer to this question. One possibility is to keep them in, and in this case each group has a small number of outliers so the effects may be comparable across groups. Another possibility is to recode the outlying cases to be missing using conditional commands outlined in Chapter 3. Another approach is to run analysis with the outliers in and the outliers excluded and to report both sets of findings. And yet another possibility is to recode television viewing from a scale variable to a categorical variable (low, medium, high viewing) and to redo the analysis using crosstabs. As we stress here and elsewhere in this book, successfully performing and interpreting statistical findings hinges on understanding the unique considerations that emerge in the context of performing analyses, and creatively adjusting to these challenges.

T Tests

A **t test** is a special case of analysis of variance that compares the means of only two categories. There are two types—an independent samples t test and a paired samples t test. An independent samples t test is equivalent to a one-way analysis of variance with two categories. In fact, you will get the exact same significance values from an independent samples t test as from a one-way ANOVA. Although it has limited applicability to the types of data included with this text, we introduce it because the t test is very popular, and in certain circumstances, serves as a better tool than ANOVA. There are also disciplinary tendencies to use t tests versus ANOVAs. Our reading of the sociological and psychological literatures suggests that sociologists are more inclined to use ANOVA, whereas psychologists tend to gravitate toward t tests.

Independent Samples T Test

Independent samples t tests operate under the same general assumptions as ANOVAs, as the goal is to compare two groups on the basis of a scale variable. An **independent samples t test** is appropriate for comparing the mean time spent watching TV for men and women. To do so, use the following commands:

Analyze
    Compare Means
        Independent-Samples T test
            Test Variables: TVHOURS
            Grouping Variable: SEX
            Define Groups: Group 1: 0 (Male)
                Group 2: 1 (Female)
                Continue
            OK

The *Define Groups* window tells SPSS the values of the two categories to be compared. It is important that you type in the values for the categories exactly as they were entered into the data set. If you typed in "M" and "F" in *Define Groups*, but they were entered into the data set as 1 and 2, SPSS would give you an error.

Independent Samples Test

		Levene's Test for Equality of Variances		t-test for Equality of Means						
									95% Confidence Interval of the Difference	
		F	Sig.	t	df	Sig. (2-tailed)	Mean Difference	Std. Error Difference	Lower	Upper
TVHOURS HOURS PER DAY WATCHING TV	Equal variances assumed	5.620	.018	1.520	897	.129	.266	.175	-.077	.609
	Equal variances not assumed			1.543	887.775	.123	.266	.172	-.072	.604

Figure 6.9 Independent Samples T test Output

The independent samples t test output in Figure 6.9 can be tricky to read. We focus here on the top row of the columns labeled "t" and "Sig. (2 tailed)" is of interest. "t" is the test statistic and Sig., once again, is the significance level. A t value close to 0 indicates that the two means are very similar and will result in a large significance value. A t value far from 0, in either a positive or negative direction, will result in a small significance level, indicating a difference in the means. This output has a t = 1.520 and a significance level = .129. Since the significance level is above .05, there is no significant association between sex and amount of time spent watching television—men and women watch about the same amount of television (2.72 and 2.99 hours per week). The Group Statistics table (Figure 6.10) shows these means.

Group Statistics

	SEX RESPONDENTS SEX	N	Mean	Std. Deviation	Std. Error Mean
TVHOURS HOURS PER DAY WATCHING TV	0 FEMALE	479	2.99	2.879	.132
	1 MALE	420	2.72	2.277	.111

Figure 6.10 Mean TVHOURS By SEX

Paired-Samples T Test

T tests also have the ability to examine observations that are not independent from one another. This occurs in a variety of circumstances. For example, consider this situation. People vary widely in their propensity for depression. For example, if we did a survey following the World Trade Center attack, it would be surprising if some people did not report high levels of depression. Yet it would be errant to attribute all of the depression to the attacks, since some people were already depressed before the attacks. If the purpose of the study is to assess how the attack *affected* people's depression levels, it would be important to take into account that some people were more prone to depression even before the attack. A researcher who was lucky enough to have measured depression levels on a sample of people before the attack could then measure those *same individual's* depression levels after the attacks. These observations would then be paired, in the sense that there are two observations for each individual. The question then becomes: is the World Trade Center attack associated with a *change in the scores of individuals'*, and does not simply reflect differences in people's overall depression levels. In situations such as this, in which the data in the two categories are related, a **paired samples t test** is appropriate.[17]

But this is not the only type of situation where a paired samples t test is appropriate. Consider this question that can be answered with the STATES07 data: Are African Americans more likely to be incarcerated than White Americans? With the structure of this data set, it is necessary to use a paired samples t test. Because each state (case) contributed data on both the incarceration rate for Blacks (PRC61) and the incarceration rate for Whites (PRC58), it is likely that these values will be correlated. For example, states with get-tough policies on crime will likely have incarceration rates higher for both African Americans and Whites. Therefore, the samples for each category are not independent, making this an appropriate situation for the paired t test illustrated below.

Analyze
Compare Means
Paired-Samples T test
Paired Variables: CRC59 (White State Prisoner Incarceration Rate per 100,000: 2005)
CRC60 (Black State Prisoner Incarceration Rate per 100,000: 2005)
(Note: Click on both variables before clicking the arrow)
OK

The *Paired Samples Statistics* table (Figure 6.11) shows that the mean incarceration rate for Whites is 408 per 100,000 people, whereas the mean incarceration rate for Blacks is 2542 per 100,000 people. The *Paired Samples Test* table (Figure 6.12) reveals that this difference is highly significant, Sig. = .000. In conclusion, there is a large association between the incarceration rate and race—the rate is much higher for African Americans than for Whites.

Paired Samples Statistics

		Mean	N	Std. Deviation	Std. Error Mean
Pair 1	CRC59 White State Prisoner Incarceration Rate per 100,000: 2005	407.898	49	150.6373	21.5196
	CRC60 Black State Prisoner Incarceration Rate per 100,000: 2005	2541.857	49	802.5104	114.6443

Figure 6.11 Sample Statistics for PRC58 and PRC61

[17] There is an ANOVA equivalent to the paired t-test called repeated measures ANOVA. It is used when there are three or more associated observations per case.

Paired Samples Test

		Paired Differences					t	df	Sig. (2-tailed)
		Mean	Std. Deviation	Std. Error Mean	95% Confidence Interval of the Difference				
					Lower	Upper			
Pair 1	CRC59 White State Prisoner Incarceration Rate per 100,000: 2005 - CRC60 Black State Prisoner Incarceration Rate per 100,000: 2005	-2133.96	766.1906	109.4558	-2354.03	-1913.88	-19.496	48	.000

Figure 6.12 Paired T Test Output

The choice to use paired samples t tests over independent samples t tests rests on the need to account for the non-independence of samples. If there is a reason to believe that the observations themselves are linked in meaningful ways (e.g., the reports of husbands being linked to the reports of wives, the incarceration of Whites being linked to the incarceration rates of African Americans, or the experiences of an individual at one point in time influencing their experiences at another point in time), paired t tests are appropriate.

Summary

Researchers can test for an association between a scale and a categorical variable by performing an ANOVA and creating bar graphs and box plots. Performing post-hoc tests further refine the results. Independent samples t tests and paired samples t tests compare the mean values of two independent or related groups. As researchers perform analyses, they should watch for characteristics of variables that influence the decision to use these methods. Some concerns include adequate sample sizes for all categories, distributions of scale variables that conform to normality and equal variance assumptions, and independent observations.

Key Terms

- Bar charts
- Box plots
- Independence assumption
- Independent Samples T test
- Normality assumption
- One-way Analysis of Variance
- Paired Samples T test
- Post-hoc tests
- Tukey test

Chapter 6 Exercises

Name_________________________________ Date_____________________

1. Astrologers assert that our birth dates influence our success (or lack thereof) in life. Test this assumption with the GSS04 data by analyzing the relationship between ZODIAC (Respondent's Astrological Sign) with SEI (Respondent's Socioeconomic Index). The SEI is an indicator of economic and social-economic attainment. The higher the SEI score, the more successful the respondent.

    Mean SEI of Pisces ____________________

    Mean SEI of Taurus ____________________

    Mean SEI of your astrological sign ____________________

    ANOVA significance level ____________________

    Is the relationship statistically significant? Yes No

    On the basis of these data, would you say that astrologers are correct in their assertion of the power of the stars? Why?

2. Test whether there is an association between a person's educational attainment and the age at which they first have children. Use the GSS04 data set to perform an ANOVA on respondents' highest educational degree (DEGREE) and their age when they had their first child (AGEKDBRN).

   Mean age for people with less than a high school degree ____________________

   Mean age for people with a high school degree ____________________

   Mean age for people with a junior college degree ____________________

   Mean age for people with a Bachelor's degree ____________________

   Mean age for people with a graduate degree ____________________

   ANOVA significance level ____________________

   Is the relationship statistically significant? Yes No

   According to the Tukey test, which categories have means that are significantly different from the means of those with a high school degree?

   On the basis of these data, would you say that education is associated with the age at which people have children? What could be the cause of this relationship?

3. Test whether there is an association between a person's race and the age at which they first have children. Use the GSS04 data set to perform an ANOVA on RACE and AGEKDBRN.

   Mean age for Whites ____________________

   Mean age for African Americans ____________________

   Mean age for other races ____________________

   ANOVA significance level ____________________

   Is the relationship statistically significant? Yes No

   According to the Tukey test, which categories have means that are significantly different from the means of Whites?

   On the basis of these data, would you say that race is associated with the age at which people have children? What could be the cause of this relationship?

4. Test whether there is an association between a person's gender and the age at which they first have children. Use the GSS04 data set to perform an independent samples t test on SEX and AGEKDBRN.

   Mean age for men ____________________

   Mean age for women ____________________

   t test equality of means significance level ____________________

   Is the relationship statistically significant? Yes No

   On the basis of these data, would you say that gender is associated with the age at which people have children? What could be the cause of this relationship?

5. Using the GSS04 data, construct and print a refined bar chart of the relationship between the age at which a person had their first child (AGEKDBRN) and one of the following categorical variables: DEGREE, RACE or SEX.

   Describe your findings below and print a refined bar chart.

6. Using the GSS04 data, identify and then test one additional scale variable that may be associated with the age at which a person has their first child (AGEKDBRN).

   Describe your findings below and print a refined box plot.

7. Using the GSS04 data, perform an independent samples t test to compare the mean socioeconomic index SEI of those who have had a born again experience with those who have not (REBORN).

Mean SEI of those non-born again ____________

Mean SEI of those born again ____________

Significance for equality of means ____________

Is the relationship statistically significant? ____Yes No___

On the basis of these data, is socioeconomic status related to the religious experience of being born again? What were you expecting and what did you find?

8. Use the GSS04 data to identify and then test one additional categorical variable that may be associated with a born again experience (REBORN).

   Describe your findings below and print a box plot

9. Perform a paired t test to compare the median earnings of male full-time workers (EMS161) to the median earnings of female full-time workers (EMS162) using the STATES07 data set.

   Mean earnings of men ____________

   Mean earnings of women ____________

   Significance for equality of means ____________

   Is the relationship statistically significant? Yes No

   Are earnings related to gender? What were you expecting and what did you find?

10. Create a new variable called WAGEGAP that is the difference between median earnings of male full-time workers (EMS161) and the median earnings of female full-time workers (EMS162). Make sure you label this variable.

    Create a histogram of WAGEGAP's distribution.

    Describe the shape of the distribution.

    In which state do women have the best earnings relative to men's? ____________________

    In which state do women have the worst earnings relative to men's? ____________________

Chapter 7:
Multivariate Analysis with Linear Regression

Overview

Chapters 5 and 6 examined methods to test relationships between two variables. Many research projects, however, require multivariate analyses to test the relationships of multiple independent variables with a dependent variable. This chapter describes why researchers use multivariate analyses and then examines one of the most powerful approaches: linear regression. We show how to interpret regression statistics and graph linear regressions using the STATES07 data. Finally, we discuss issues related to data structures and model building.

The Advantages of Multivariate Analysis

In most studies, multivariate analysis is the final stage of data analysis. In the first stage, researchers use univariate analyses to understand variables' structures and distributions, which affect the choice of what multivariate analyses techniques to use. In the second stage, researchers use bivariate analyses to understand relationships between variable pairs, which in multivariate analyses can "disappear" or emerge in new—and sometimes unexpected—ways. As we discuss later in this chapter, a core concern in multivariate analysis is creating robust models that satisfy both mathematical and theoretical assumptions. A firm understanding of the data at the individual and bivariate levels is critical to satisfying this goal.

Before discussing model building and the application of multiple linear regression, let us first take a step back and reflect on the reasons why multivariate analyses are performed. Suppose, for example, that a researcher is interested in predicting academic success, a construct that she operationalizes as grade point averages (GPAs). After determining that GPAs are normally distributed, she performs a series of bivariate analyses that reveal the following sets of relationships:

- Women have higher GPAs than men
- Length of time spent studying is positively associated with GPAs
- Students who are members of fraternities have the same GPAs as non members
- Students who are in sport teams have lower GPAs than non-athletes
- Sophomores, juniors, and seniors have higher GPAs than freshmen
- Students who drink heavily have lower GPAs than light drinkers and abstainers

These are interesting findings, but they open new questions. For example, do freshman and athletes receive lower grades because they study less? Are the GPAs of men pulled downward because they are more likely to be members of sports teams? Can the differences between the performances of men and women be attributed to men drinking more heavily? Answering these types of questions requires considering not only the relationships between the dependent variable (GPA) and individual variables (gender, drinking, sports, etc.), but also the constellation of variables that correspond with being a student.

One of the great benefits of multivariate analysis is its ability to document **collective effects**—the interplay among factors on predicted outcomes. For instance, multivariate analyses can predict the expected GPAs based on combinations of variables as they may be configured in the lives of individuals (e.g., an upper class, non-drinking, female athlete). They also can assess the amount of variation in the dependent variable that can be attributed to the variables in the model, and conversely, how much of the variation is left unexplained. In the case of GPAs, a multivariate model can tell the strength of the six factors in predicting academic success. Do these factors account for many of the differences among students, or only a small fraction of the differences? Multivariate analysis measure **explanatory power**, and how well predictions of social behavior correspond with observations of social behavior.

Additionally, multivariate analyses are critical to accounting for the potential impact of **spurious relationships**. Recall from Chapter 1 that a spurious relationship occurs when a third variable creates the appearance of relationship between two other variables, but this relationship disappears when that third variable is included in the analysis. Using the example above, perhaps the differences in performances of athletes and non-athletes may not be due to sports, but may simply be the result of athletes spending less time studying. If the negative association of athletics to GPA "disappears" when studying is taken into account, it leads to a more sophisticated understanding of social behavior, and more informed policy recommendations.

Finally, one of the great advantages of multivariate models is that they allow for the inclusion of **control variables**. Control variables not only help researchers account for spurious relationships, they gauge the impact of any given variable above and beyond the effects of other variables. For example, a researcher could document the influence of drinking on GPAs adjusting for the impact of gender, sports, fraternities, and time spent studying. Or consider the relationship between gender and GPA. Suppose the relationship between gender and GPA disappears after taking into account all of the other variables in the model. What would that suggest about theories that posit innate differences in abilities to succeed in college?

Putting Theory First—When to Pursue Linear Regression

Later in this chapter we consider some mathematical principles and assumptions that underpin linear regression. But first we consider some theoretical issues critical to its application. Linear regressions are designed to measure one specific type of relationship between variables: those that take **linear** form. The theoretical assumption is that for every one-unit change in the independent variable, there will be a consistent and uniform change in the dependent variable. Perhaps one reason why linear regression is so popular is that this is a fairly easy way to conceive of social behavior – if more of one thing is added, the other thing will increase or decrease proportionately. In fact, many types of relationships operate this way. More calories results in proportional weight gains, more education results in proportionally higher earnings, etc. In our example above, a linear model assumes that that each additional hour a student spends studying (whether the increase is from 5 to 6 hours a day, or from 1 to 2 hours a day) the incremental effect on the GPA will be constant. This may be true, but it also may not.

Recall the discussion in Chapter 5, that there are many types of relationships between variables. While many sets of relationships are linear, many others take on very different forms. For example, students who experience little anxiety and those who experience excessive anxiety tend to perform more poorly on exams than students who score midrange in an anxiety scale (these individuals are very alert, but not overwhelmed). Because this "inverted U" shaped relationship is nonlinear, the application of linear techniques will make it appear non-existent. Or consider another example - the impact of class size on academic performance. It is generally understood that the smaller the class the more students benefit, thus there is a negative relationship between class size and academic performance. However, changing a class from 25 students to 20 students will have almost no effect on changing the way a classroom operates. In contrast, the same increment of change from 12 to 7 students can have a much more substantial change in classroom dynamics.[18] Conversely, there may be positive effects of time spent studying on GPA, but the benefits of each additional hour may be smaller once a sufficient amount of study time is reached.

Linear Regression: A Bivariate Example

Later in this chapter we detail some criteria for applying linear regression. But before bogging down the discussion in these cautions, let us look at its application and consider how regression statistics are interpreted. Note, though, that the structure of the dependent variable is a critical consideration, as linear regressions are performed on scale dependent variables. In our sample illustration, we will be testing the relationship between poverty (independent variable) and the teenage birth rate (dependent variable) using the STATES07 data. Our guiding hypothesis is that poverty is associated with teen births and that places with higher poverty rates likely have higher teen birth rates. Before considering why this relationship may exist, we determine *if* it exists.

We already identified some approaches to examining relationships between two scale variables in Chapter 5 - correlations and scatter plots. The scatter plot of these variables in Figure 7.1 shows that the data points tend to flow from the lower left-hand corner of the graph to the upper right. The correlation of these two variables is .774, a strong positive relationship. The linear regression determines the equation of the line that best describes that relationship. This equation can be used to predict values of the dependent variable from values of the independent variable.

---

[18] This type of *logarithmic* relationship can still be tested using linear regression techniques, but it requires transforming data so that the model corresponds to the way the data are actually configured. Describing these transformations is beyond the scope of this book, but for a description of these methods, see *Applied Linear Statistical Models* by Michael H. Kutner, John Neter, Christopher J. Nachtsheim, and William Wasserman (2004).

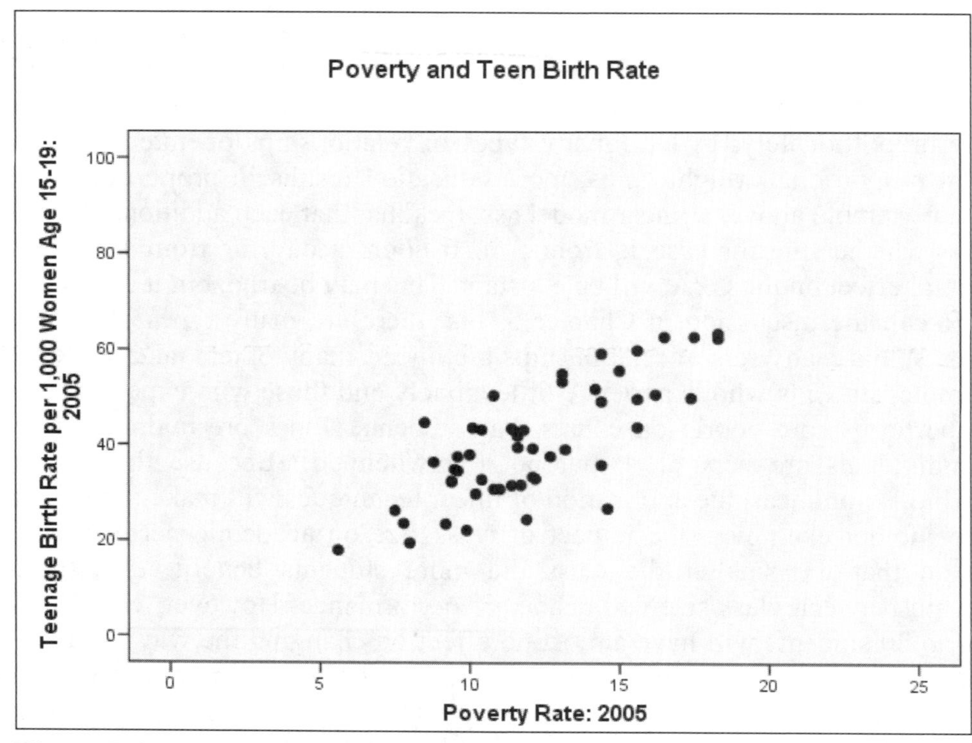

Figure 7.1 Scatter Plot of PVS493 and DMS380

To perform a regression, open the STATES07 data and specify the following variables in the regression menu:

Analyze
Regression
Linear
Dependent: DMS380 (Teenage Birth Rate: 2005)
Independent(s): PVS493 (Poverty Rate: 2005)
OK

Figure 7.2 contains the resulting regression output. We will concentrate on three groups of statistics from this output: the coefficients, the significance tests, and the R square statistic.

Interpreting Linear Regression Coefficients

The **unstandardized coefficient** of an independent variable (also called **B** or **slope**) measures the strength of its relationship with the dependent variable. It is interpreted as the size of the difference in the dependent variable that corresponds with a one-unit difference in the independent variable. A **coefficient** of 0 means that the values of the dependent variable do not consistently differ as the values of the independent variable increase. In that case, we would conclude that there is no linear relationship between the variables.

In our model, the coefficient for poverty rate is 3.170. For every one-percentage increase in the poverty rate, there is a predicted increase in the teen birth rate of more than three births (3.170) per 1,000 teenage women.

Model Summary

Model	R	R Square	Adjusted R Square	Std. Error of the Estimate
1	.774[a]	.599	.591	7.768

a. Predictors: (Constant), PVS493 Poverty Rate: 2005

ANOVA[b]

Model		Sum of Squares	df	Mean Square	F	Sig.
1	Regression	4418.244	1	4418.244	73.227	.000[a]
	Residual	2956.464	49	60.336		
	Total	7374.708	50			

a. Predictors: (Constant), PVS493 Poverty Rate: 2005

b. Dependent Variable: DMS380 Teenage Birth Rate per 1,000 Women Age 15-19: 2005

Coefficients [a]

		Unstandardized Coefficients		Standardized Coefficients		
Model		B	Std. Error	Beta	t	Sig.
1	(Constant)	2.125	4.604		.462	.646
	PVS493 Poverty Rate: 2005	3.170	.370	.774	8.557	.000

a Dependent Variable: DMS380 Teenage Birth Rate per 1,000 Women Age 15-19: 2005

Figure 7.2 Regression Output

Moving across the row for "Poverty Rate: 2005" in the Coefficients Table, we find a significance **(Sig.) score**. The significance score of .000 indicates that chance is an extremely unlikely explanation, as there is less than a 1/1,000 chance of a relationship this strong emerging, within a data set this large simply because of random chance. Because the relationship is significant, we are confident of an actual association between poverty and teen births.

The B column also shows the **constant**, a statistic indicating the **intercept**—the predicted value of the dependent variable when the independent variable has a value of 0. In this example, the constant is the predicted number of births per 1,000 teenage women if a state had no one living below the poverty line (a poverty rate of 0). Even if a state had no poor people, we could still expect about two births (2.125) each year for every 1,000 teenage women. The intercept also has a significance level associated with it, but this statistic is usually ignored. Below we will show how the constant is used in predictions and the formulation of a regression line.

Interpreting the R-Square Statistic

The R-square measures the regression model's usefulness in predicting outcomes—indicating how much of the dependent variable's variation is due to its relationship with the independent variable(s). An R-square of 1 means that the independent variable explains 100% of the dependent variable's variation—it entirely determines its values. Conversely, an R-square of 0 means that the independent variable explains none of the variation in the dependent variable and it has no explanatory power whatsoever.

The Model Summary Table for our example shows the R-square is .599, meaning 59.9% of the variation from state to state in the teenage birth rate can be explained by their poverty rates. The remaining 40.1% can be explained by other factors that are not in the model. You may have also noticed that next to the R-square statistic is correlation of the two variables, .774. In bivariate linear regressions, the R-square is actually calculated by squaring the correlation coefficient (.774*.774=.599).

Putting the Statistics Together

In sum, there are three important statistics to attend to in a linear regression model. First, the regression coefficients measure the strength and direction of the relationships. Second, for each of these regression coefficients, there is a significance score, which measures the likelihood that the relationship revealed in the coefficients can be attributed to random chance. Finally, the R-square statistic measures the model's overall predictive power and the extent to which the variables in the model explain the variation in the dependent variable.

In our example, we found that teenage births are positively related to poverty—teenage girls living in poorer states have, on average, more children than those in more affluent states. We also observed that this relationship could not be attributed to the effects of random chance. Finally, we discovered that poverty has very strong predictive powers, as this one variable accounts for nearly two thirds (59.9%) of the variation in teen births within the United States.

Using Linear Regression Coefficients to Make Predictions

One of the most useful applications of linear regression coefficients is to make "what if" predictions. For example, what if we were able to reduce the number of families living in poverty by a given percentage? What effect would that have on the frequency of teen births? One way to generate answers to these questions is to use regression formulas to predict values for a dependent variable:

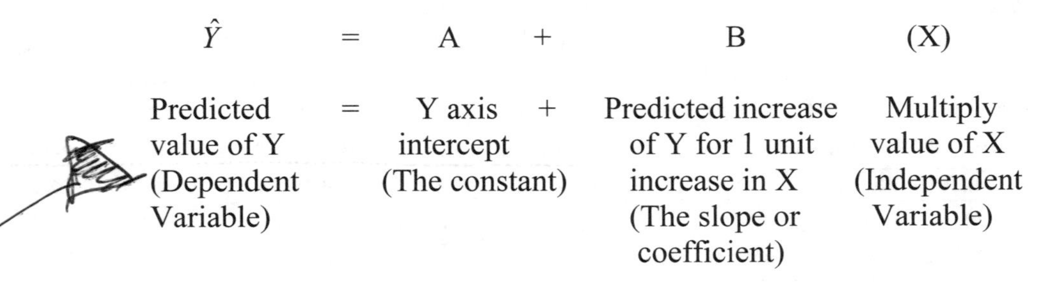

$\hat{Y}$	=	A	+	B	(X)
Predicted value of Y (Dependent Variable)	=	Y axis intercept (The constant)	+	Predicted increase of Y for 1 unit increase in X (The slope or coefficient)	Multiply value of X (Independent Variable)

This equation, combined with the information provided by the regression output, allows researchers to predict the value of the dependent variable for any value of the independent variable. Suppose, for example, that we wanted to predict the teenage birth rate if the poverty rate is 20%. To do this, substitute 20 for X in the regression equation and the values for the constant and coefficient from the regression output.

$\hat{Y}$	=	A	+	B	(X)
Predicted	=	2.125	+	3.170	(?)

$\hat{Y}$	=	A	+	B	(X)

65.53	=	2.125	+	3.170	(20)

We calculate that a state with a poverty rate of 20% will have a predicted teenage birth rate of 65.53, approximately 65 births per 1,000 teenage women.

Making predictions from regression coefficients can help gauge the effects of social policy. We can predict how much the teenage birth rate *could* decline if poverty rates in states were reduced. What would the teenage birth rate to be if a state could reduce its poverty rate from 15% to 10%?

$\hat{Y}$	=	A	+	B	(X)
49.68	=	2.125	+	3.170	(15)

$\hat{Y}$	=	A	+	B	(X)
33.83	=	2.125	+	3.170	(10)

Teen births at 15% poverty rate = 49.68/1000
Teen births at 10% poverty rate = 33.83/1000
Reduction in teen births = 15.85/1000 teenage women

It is good practice only to use values in the independent variable's available range when making predictions. Because we used data with poverty rates between 1% and 22% to construct the regression equation, we predict teen pregnancy rates only for poverty rates within this range. The relationship could change for poverty rates beyond 22%. It could level off, or even decrease, or the rates could skyrocket, as some sociological studies indicate. Because our data do not tell us about the relationship for places of concentrated poverty, we must not use the regression line to make predictions about them.

Using Coefficients to Graph Bivariate Regression Lines

Regression coefficients are generally not meaningful to audiences unfamiliar with statistics, but graphs of regression lines are. To illustrate, we create a regression line using the independent variable, PVS493, the coefficients, and the predicted values of the dependent variable. To make the predicted values, we will compute a new variable based on information in the regression output.

Transform
    Compute Variable
        Target Variable: BIRTHPRE
        Numeric Expression: 2.125 + 3.170*PVS493
        Type and Label
            Label: Predicted Teen Birth Rate From Poverty Rates
            Continue
        OK

Note: BIRTHPRE = predicted teen birth rate
2.125 = constant
3.170 = coefficient for poverty rates
PVS493 = poverty rates for states

By creating the new variable BIRTHPRE, we are able to construct a new visual depiction of the data. In this case, we want to create a scatter plot of the teen birth rate (DMS380) and the poverty rate (PVS493). We also want to superimpose a line on that scatter plot that represents the predicted teen birth rates (BIRTHPRE) as related to poverty rates (PVS493). To do this, we are going to create a single graph that has two pairings of variables (DMS380-PVS493 and BIRTHPRE-PVS493). This is one of the functions of the *SPSS Chart Builder* (which can also facilitate the construction of other types of graphs as well). Below we discuss both how to construct this graph and the general ways the *Chart Builder* works.

First locate the *Chart Builder* window with the following command:

Graphs
Chart Builder

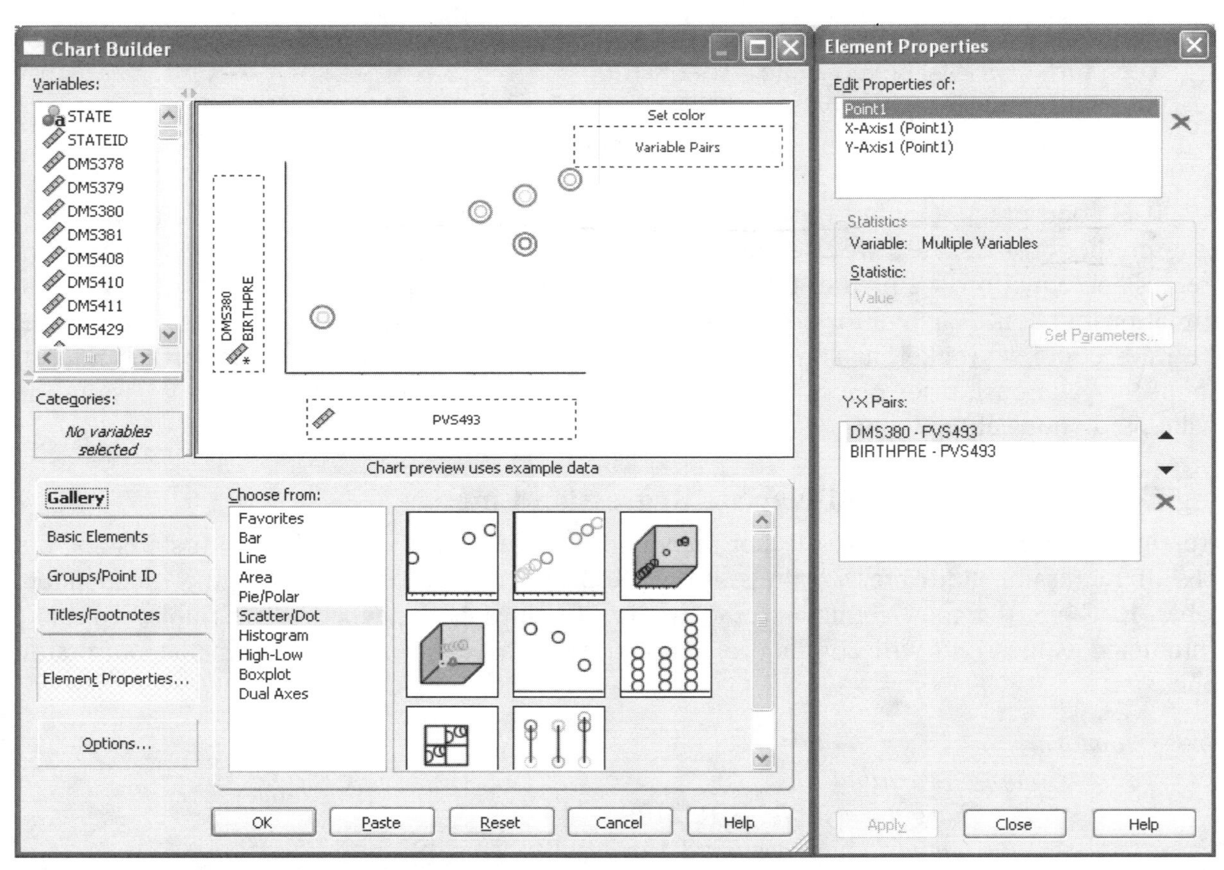

Figure 7.3 Chart Builder Dialog Windows

Your screen should look largely like Figure 7.3, but will be lacking some of the specifications that we have already entered. First note the various options in the *Chart Builder*, such as bar charts, line charts, box plots. And once a classification of charts is selected, a variety of different chart options emerge, such as matrixes, three dimensional plots, etc. In this case, you will want to select "Scatter/Dot" from the *Choose from:* dialog box.

Next we want to specify the variables. Recall that in scatter plots, the convention is for the dependent variable to be arrayed on the Y axis and the independent variable to be arrayed on the X axis. In this case, our model asserts that poverty influences teen birth rates, thus poverty rates (PVS493) is placed on the X axis. To do this, drag PVS493 from *Variables:* to its desired location on the X axis. Then to move DMS380 and BIRTHPRE, highlight both variables by holding down the Ctrl key, and again drag them into the desired location on the Y axis. You should see all three of the variables listed, as shown in Figure 7.3. Note also that the box in the far lower left shows that our desired Y-X pairs have been created.

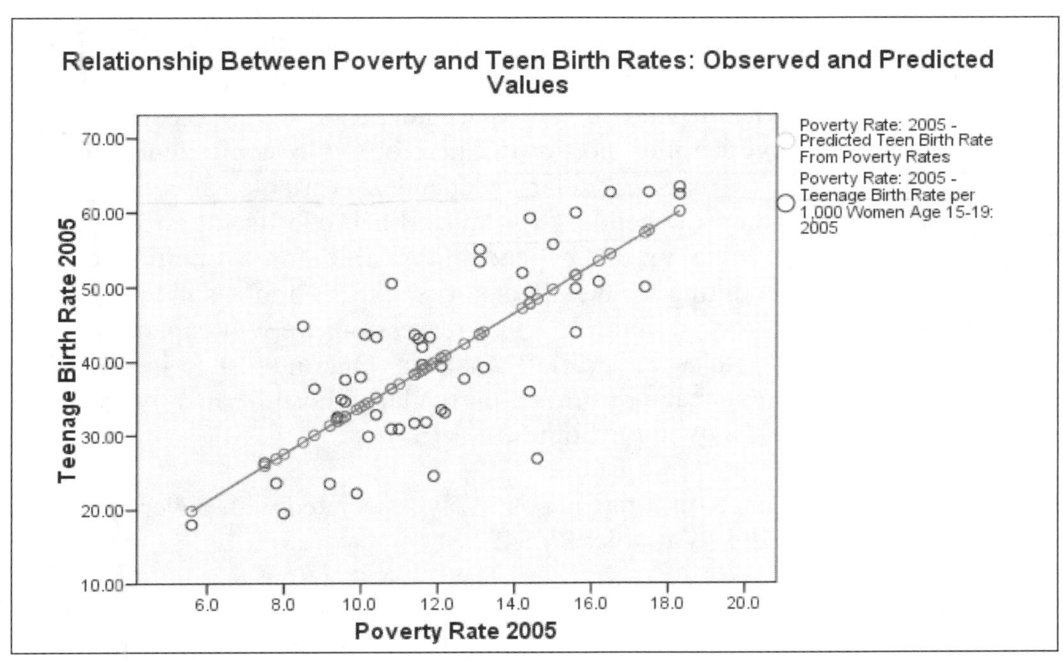

Figure 7.4 An Overlay Scatterplot of Observed and Predicted Values

After you run these commands, you will observe that SPSS creates a chart that looks fairly similar to the one represented in Figure 7.4. We modified this graph, however, by inserting a title, axis labels, and a trend line. Try to replicate this graph. In most instances, you will be able to make the desired changes by double clicking on the chart to open the *Chart Editor* window, and right clicking on the elements of the chart that you wish to modify. Notice, for example, that we imposed an interpolation line on the predicted values by first selecting those values and then by choosing that option with a right click of the mouse. It will likely take some experimentation to master the process of making graphs with the *Chart Builder*.

Multiple Linear Regression

In a multiple linear regression, more than one independent variable is included in the regression model. As we discussed earlier, multiple regression examines how two or more variables act together to affect the dependent variable. This allows researchers to introduce control variables that may account for observed relationships, as well as document cumulative effects.

While the interpretation of the statistics in multiple regression is, on the whole, the same as in bivariate regression, there is one important difference. In bivariate regression, the regression coefficient is interpreted as the predicted change in the value of the dependent variable for a one-unit change in the independent variable. In multiple regression, the effects of multiple independent variables overlap in their association with the dependent variable. The coefficients printed by SPSS don't include the overlapping part of the association—only the unique association between the dependent and that independent variable (but the R-square statistic does include this overlapping portion). Therefore, the coefficient shows the "net strength" of the relationship of that particular independent variable, above and beyond the relationships maintained by other independent variables. Each coefficient is then interpreted as the predicted change in the value of the dependent variable for a one-unit change in the independent variable, after accounting for the effects of the other variables in the model.

To illustrate how to perform a multiple linear regression, we will expand the study of the teenage birth rate (DMS380) into a multivariate analysis. Our interest is in identifying some factors that may relate with the teenage birth rate, particularly those frequently forwarded in the popular press. We will test the following hypotheses:

Hypothesis 1: The teenage birth rate is positively associated with poverty.
Independent Variable: PVS493

Hypothesis 2: The teenage birth rate is negatively associated with per capita spending for education.
Independent Variable: EDS133

Hypothesis 3: The teenage birth rate is positively associated with the amount of welfare (TANF)[19] families receive.
Independent Variable: PVS522

Hypothesis 4: Teen Birth Rate is positively associated with the percent of the population that is African American
Independent Variable: DMS441

To run this regression, and produce the output reproduced in Figure 7.5, use the following commands:

[19] TANF stands for "Temporary Assistance to Needy Families," the most commonly understood program equated with "welfare."

Analyze
Regression
Linear
Dependent: DMS380 (Teenage Birth Rate: 2005)
Independents: PVS493 (Poverty Rate: 2005)
EDS133 (Expenditures per Pupil in 2004)
PVS522 (Average Monthly TANF Assistance per Recipient: 2004)
DMS441 (Percent of Population Black 2005)
OK

Model Summary[b]

Model	R	R Square	Adjusted R Square	Std. Error of the Estimate
1	.852[a]	.726	.703	6.623

a. Predictors: (Constant), DMS441 Percent of Population Black: 2005, EDS133 Per Capita State and Local Govt. Spending for Education: 2004, PVS522 Average Monthly TANF Assistance per Recipient: 2004, PVS493 Poverty Rate: 2005

b. Dependent Variable: DMS380 Teenage Birth Rate per 1,000 Women Age 15-19: 2005

ANOVA[b]

Model		Sum of Squares	df	Mean Square	F	Sig.
1	Regression	5356.685	4	1339.171	30.526	.000[a]
	Residual	2018.023	46	43.870		
	Total	7374.708	50			

a. Predictors: (Constant), DMS441 Percent of Population Black: 2005, EDS133 Per Capita State and Local Govt. Spending for Education: 2004, PVS522 Average Monthly TANF Assistance per Recipient: 2004, PVS493 Poverty Rate: 2005

b. Dependent Variable: DMS380 Teenage Birth Rate per 1,000 Women Age 15-19: 2005

Coefficients[a]

Model		Unstandardized Coefficients		Standardized Coefficients	t	Sig.
		B	Std. Error	Beta		
1	(Constant)	26.588	9.353		2.843	.007
	PVS493 Poverty Rate: 2005	2.336	.396	.570	5.893	.000
	EDS133 Per Capita State and Local Govt. Spending for Education: 2004	-.001	.003	-.031	-.349	.728
	PVS522 Average Monthly TANF Assistance per Recipient: 2004	-.090	.023	-.373	-3.952	.000
	DMS441 Percent of Population Black: 2005	.058	.096	.055	.604	.549

a. Dependent Variable: DMS380 Teenage Birth Rate per 1,000 Women Age 15-19: 2005

Figure 7.5 Multiple Regression Output

Interpreting Multiple Linear Regression Coefficients

As in the analysis of bivariate regressions, we approach this output considering three questions. First, what are the natures of the relationships? Second, are the relationships statistically significant? And third, how powerful is the model in explaining the variation in teen birth rates within the United States?

Hypothesis 1 is supported, re-establishing the relationship identified in the previous bivariate analysis—states that have higher poverty rates tend to have higher teenage birth rates. The regression coefficient is positive (2.336) (indicating that the more poverty, the higher the teen birth rate) and the relationship is statistically significant (Sig.=.000). You may notice that the coefficient is smaller than it was in the bivariate regression model (3.170). This is the result of the multivariate model documenting the *unique* effect of poverty rates on teenage birth rates, after accounting for the other variables in the model.

Hypothesis 2 predicts that the more a state spends per pupil, the lower the teenage birth rate will be. Although the regression coefficient is negative (-.001), the relationship is not statistically significant (Sig. = .728). Here we find no support for a commonly espoused liberal thesis that allocating more money into education will necessarily result in lower teen birth rates.

But this does not necessarily indicate that conservative theories are supported, either. Hypothesis 3 predicts that the more a state spends on welfare per recipient family, the higher the teenage birth rate will be (this hypothesis tests the conservative thesis that welfare encourages irresponsible behavior). This relationship is statistically significant (Sig.= .000). However, the negative regression coefficient (-.090) shows a relationship *opposite* the one predicted in the hypothesis. The more a state spends on welfare per recipient family *the lower the teenage birth rate will be.* For every $1 a family receives in TANF payments each month, the teenage birth rate decreases by .09. If we multiply this times 250, we can predict how much an increase in $250 expenditures per welfare family per month will "buy" in lowering teenage birth rates. Controlling for all other factors, an increase in TANF payments of $250 could be expected to result in 22.5 fewer births per 1000 teenage women.

Hypothesis 4 predicts that the greater the proportion of the population that is African American, the higher the teenage birth rate. If you are curious, try a bivariate linear regression and you will indeed find a statistically significant positive relationship between these two factors. However, when we enter the other variables in a multivariate model, notice that although there is a positive regression coefficient (.058), it is not significant. Thus, this model suggests that once issues such as poverty and spending on public assistance to the poor are taken into account, the impact of race seems to disappear! This suggests that teen births may have less to do with the issue of race than it does with issues of poverty and aid to the poor.

Finally, we turn to the question of model strength. In multivariate regressions, the **adjusted R-square** statistic is used instead of the R-square because adding even unfounded independent variables to a model will raise the R-square statistic. The adjusted R-square statistic compensates for the number of variables in the model and it will only increase if added variables contribute significantly to the model. The adjusted R-square is often used to compare which of several models is best. How good is the model? The Adjusted R-square statistic in the Model Summary Table means that 70.3% (Adj R-square=.703) of the variation in the teenage birth rate can be attributed to these four variables! This is an excellent model. In fact, it is rare to find a model for social behavior that has such a high explanatory power.

In sum, the model suggests that higher teenage births are strongly predicted by poverty and lower supports to the poor. It also suggests that educational expenditures per pupil has no observable effect on teen birth rates. We also found that the relationship between race and teen births may be spurious, and this apparent relationship disappears when poverty rates are taken into account. However, it must be emphasized our intent here is not to offer a comprehensive analysis of this challenging research question, but rather to use the question to illustrate a statistical technique. Certainly, additional research is warranted to focus on a variety of related questions concerning causality (does poverty cause increased teen births rates, or do teen births influence the poverty rates?), levels of analysis (should the analysis be on state level comparisons as we do here, or on individuals?), and measurement (even if overall education funding has no effect on stemming teen births, it may matter how funds are spent).

Graphing a Multiple Regression

Graphing relationships from multiple regressions is more complex than graphing relationships from bivariate regressions, although the approach is the same. Because there are many variables in the multivariate models, the two-dimensional graphs needs to control for the effects of other variables. To graph a regression line from a multivariate model requires selecting one independent variable (X axis variable), which will be linked to the dependent variable (Y axis variable). The rest of the variables in the model will be held constant. To hold these values as constants, any value could be chosen, but the most common choice is the mean (which we generate using *Descriptives*) for scale variables and the mode for categorical variables. Again, we will use the regression formula, now expanding it to include the other variables in the model.

$$\hat{Y} = A + B_1(X_1) + B_2(X_2) + B_3(X_3) + B_4(X_4)$$

$\hat{Y}$ = Predicted Value of the dependent variable
A = Constant

B_1 = Slope of Variable 1	X_1 = Chosen value of Variable 1
B_2 = Slope of Variable 2	X_2 = Chosen value of Variable 2
B_3 = Slope of Variable 3	X_3 = Chosen value of Variable 3
B_4 = Slope of Variable 4	X_4 = Chosen value of Variable 4

This example will show how to graph the association of welfare benefits and teenage birth rates, holding poverty rates and school expenditures at their means, and race at its mode. This requires computing a new variable (TEENPRE), the predicted value of teen birth rate. Input the following equation into the *Compute* command to generate the new variable TEENPRE and label the variable "Predicted Teenage Birth Rate."

Transform
 Compute
  Target Variable: TEENPRE (type this in)
  Numeric Expression: 26.588 + (2.336*12.08) + (-.001*2231.57) + (-.090*PVS522) + (.058*11.28)
  Type and Label
   Label: Predicted Teenage Birth Rate
   Continue
  OK

Sources of numbers in the above equation:

Constant (A) = 26.588

Variable	B	Mean Value
PVS493	2.336	12.08
EDS133	- .001	2231.57
PVS522	- .090	139.77
DMS441	.058	11.28

To graph this relationship, use the *Scatter/Dot* functions in the *Graph Chart Builder* (Figure 7.6).

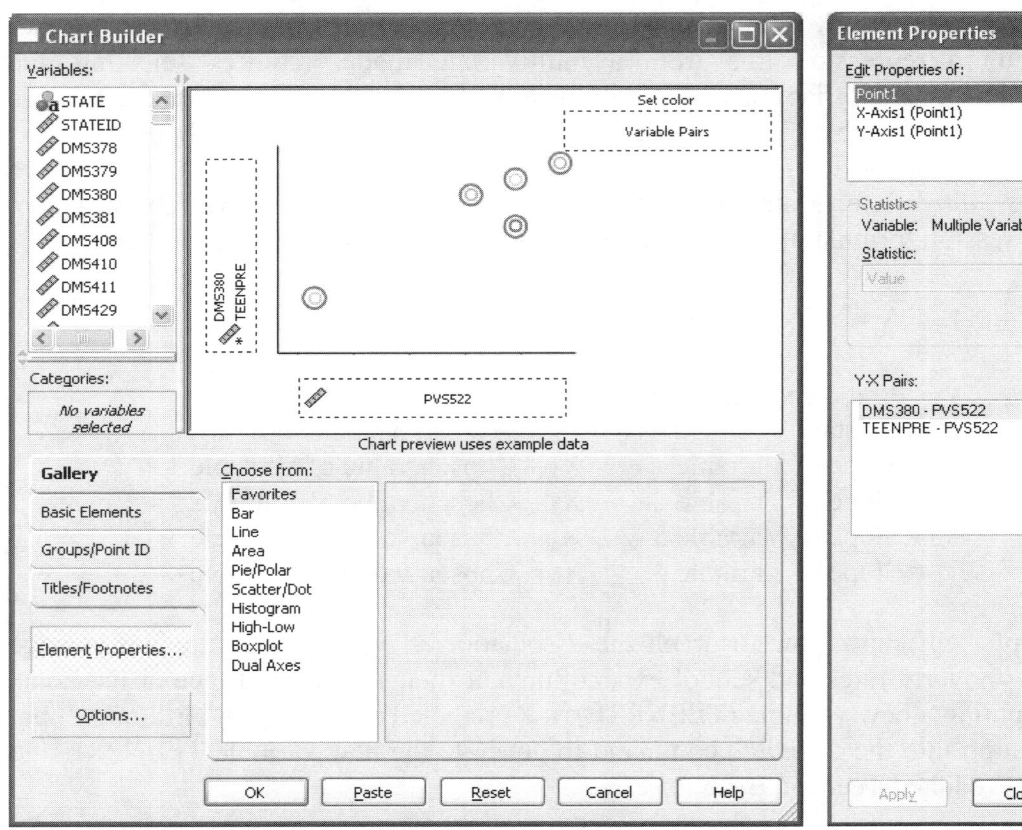

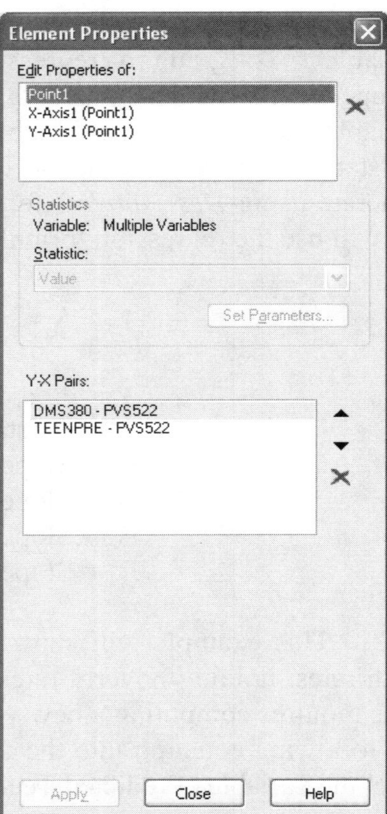

Figure 7.6 Scatter Plot Dialog Box

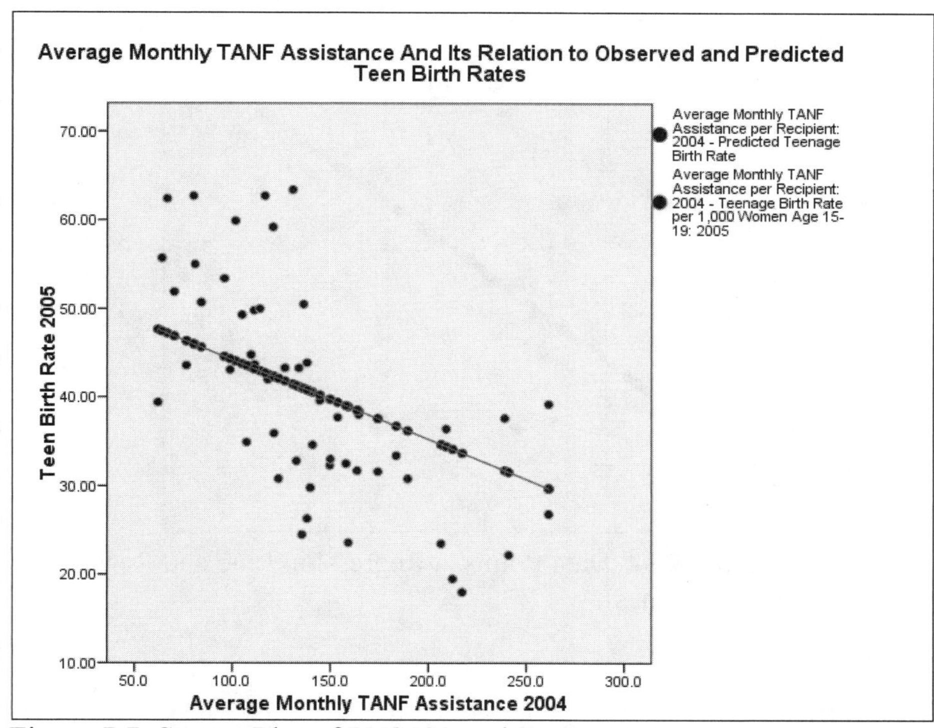

Figure 7.7 Scatter Plot of PVS522 and TEENPRE

Other Concerns in Applying Linear Regression

Just like ANOVA, linear regression has assumptions concerning the origin and structure of the dependent variable. Linear regression results are only meaningful if these assumptions have been met. Linear regression assumes that the residuals follow a normal distribution with constant standard deviation, as outlined below.

Residuals

The coefficients and significance values that result from regression analysis are calculated under the assumption that a straight line is a good model for the relationship. How well a line serves as a model for the relationship can be checked by looking at how the actual observations sit in relation to the predicted values along the line. This is measured by the vertical distance from the prediction to the actual observation and is called the **residual** (illustrated with the dashed lines in Figure 7.8). Regression coefficients are calculated so that the resulting line has the lowest possible accumulation of residuals, minimizing the overall distance between the observations and the predictions.

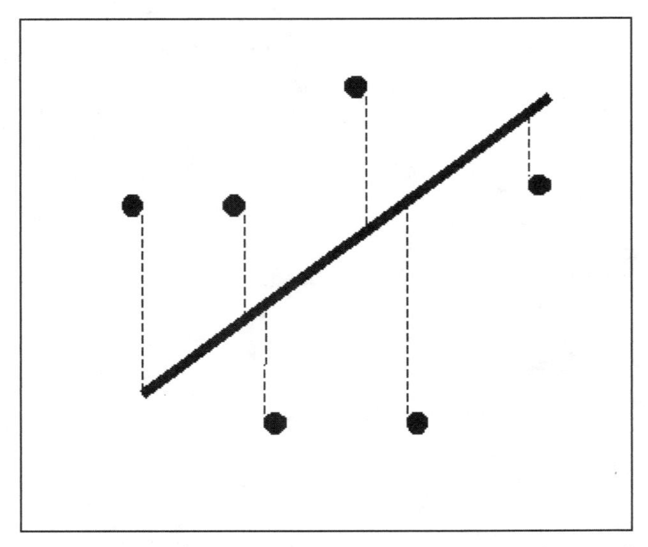

Figure 7.8 Data Points, a Regression Line and Residuals (dashed lines)

Constant Variation

In some circumstances the regression line is not well suited to the data, as in Figure 7.9, which shows a "fan effect." This regression line is much better at predicting low values of the dependent variable than it is at predicting high values, and is therefore not an appropriate model, regardless of the strength of the coefficients or their significance values. In a well-fitting regression model, the variation around the line is constant all along the line (Figure 7.3 is a good example).

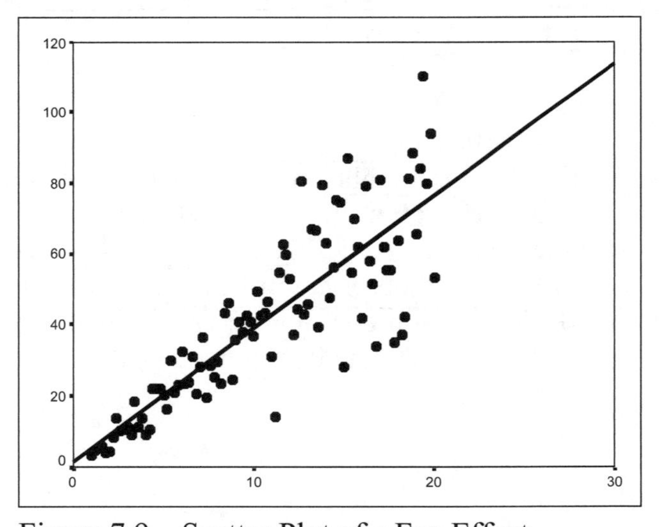

Figure 7.9 Scatter Plot of a Fan Effect

Normality of Residuals

The residuals should follow a normal distribution, with a mean of 0. Recall that a normal distribution is shaped like a bell—it is symmetric, and most points are in the middle, with fewer and fewer farther from the mean. Since the residuals measure where the points fall in relation to the line, a symmetric distribution of the residuals indicates that the same number of points fall above and below the line. Since a residual of 0 means a point is right on the line, a mean of 0 indicates the line is in the middle of the points. Once again, some are above and some are below. And the bell shape means that most are close to the line, and there are fewer points farther from the line.

One way to check the normality of residuals is to save and plot residuals from the regression command. To do so for our multivariate example:

Analyze
  Regression
    Linear
      Dependent: DMS380 (Teenage birth rate: 2005)
      Independents: PVS493 (Poverty Rate: 2005)
        EDS133 (Expenditures per Pupil in 2004)
        PVS522 (Maximum Monthly TANF Benefit for Family of Three in 2004)
        DMS441 (Percent of Population Black: 2005)

This will create a new variable RES_1, which can be graphed using the Graph command (Figure 7.10).

Graphs
  Interactive
    Histogram
      Assign Variables
        $count: RES_1
      Histogram
        Display Normal Curve
      OK

As Figure 7.10 shows, the residuals form a reasonably normal distribution, which is a good indication that the regression is working well.

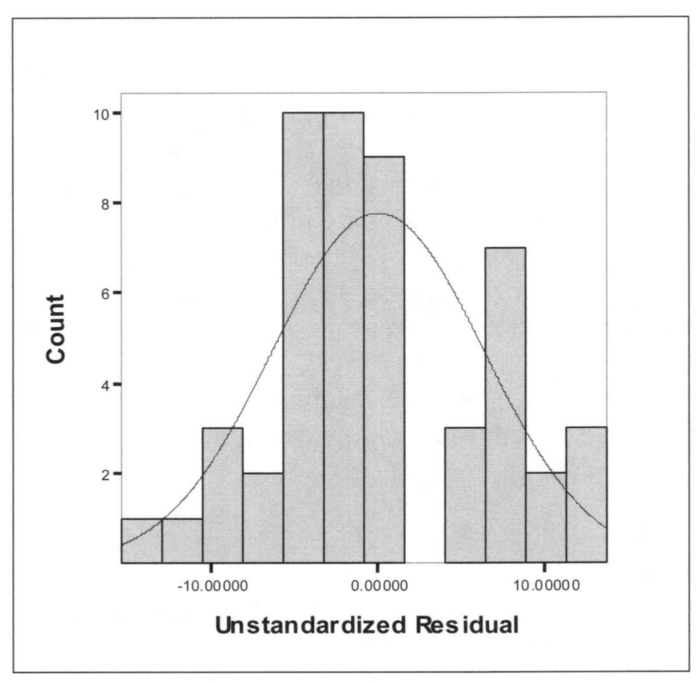

Figure 7.10 Histogram of Normally Distributed Residuals

Building Multivariate Models

How does a researcher decide which independent variables to include in a model? One seemingly efficient way is to load all of the variables in the data set into the model and see which create significant findings. Although this approach seems appealing (albeit lazy), it suffers from a number of problems.

Degrees of Freedom

The first problem is degrees of freedom. **Degrees of freedom** refer to the number of observations in a sample that are "free to vary." It is a way of compensating for the possibility of making inaccurate conclusions. Every observation increases degrees of freedom by one, but every coefficient the model estimates (including the constant) decreases the degrees of freedom by one. Because every independent variable introduced into a regression model lowers the degrees of freedom, it reduces the test's ability to find significant effects. Therefore, it is in the researcher's interest to be highly selective in choosing which variables to include in a model.

The following strategies can make decisions easier. First, researchers should select only variables for which there is a theoretical basis for inclusion. Then they should explore the data with univariate and bivariate analyses, and only include variables that have potentially interesting results, or which are needed to serve as controls. In large models, we also suggest introducing new variables sequentially. Rather than including all of the variables at once, include a small group before introducing subsequent variables. As these variables are entered into the model, observe not only how they operate, but also how the coefficients and significance scores change for the other variables as well.

Collinearity

Collinearity occurs when two or more independent variables contain strongly redundant information. If variables are collinear, there is not enough distinct information in these variables for the multiple regression to operate correctly. A multiple regression with two independent variables that measure essentially the same thing will produce errant results. An example is the poverty rate (PVS493) and the percent of children living in poverty (PVS495). These are two different variables, but they are so strongly collinear (correlation of .94) that they are nearly indistinguishable in the regression equation.

Collinear variables create peculiar regression results. For example, the two correlated poverty variables have significant bivariate relationships with teenage birthrate. However, if we run a multiple regression (you can try this) of PVS493 and PVS495 with teenage birth rates (DMS380), *both* variables become *insignificant*. Unlike the issue of spurious relationships, here the disappearance of the relationship is not explained away, but is rather the result of a mathematical corruption of the model.

The implication is that researchers should be careful about putting highly correlated variables into regression equations. To check for collinearity, start by examining a correlation matrix that compares all independent variables with each other.[20] A correlation coefficient above .50 is an indicator that collinearity *might* be present. If it is, variables may need to be analyzed in separate regression equations, and then how they operate together when included in the same model. It is only fair to mention, however, that collinearity this extreme is quite rare.

Dummy Variables

Recall that the coefficients for independent variables reflect incremental differences in the dependent variables. This means the values of independent variables need to have meaningful increments. This works easily for scale variables, but not categorical ones. Categorical independent variables, such as race, can be made to have meaningful increments through **dummy coding** (also called indicator coding). Each dummy variable splits the observations into two categories. One category is coded "1" and the other "0." With this coding, a one-unit difference is always the difference between 1 and 0, the difference of being in one category compared to the other. The coefficient is then interpreted as the average difference in the dependent variable between the two categories. Dummy variables are always coded 1 and 0. Although values of "1" and "2" seem logical and are also one unit apart, there are mathematical conveniences that make "1" and "0" preferable.

Dummy coding is easy for a variable with two categories, such as employment status: (employed vs. unemployed), or gender (men vs. women). Consider this example. Suppose we wanted to assess who has saved more for retirement, employed people or unemployed people. Since we are comparing two groups, the employed and the unemployed, employed would be coded as 1 and unemployed would be coded as 0. If the coefficient for employment status were 4,600, it would mean that in general, employed people had saved an average of $4,600 more than unemployed people.

Some categorical variables have more than two categories, such as race, or geographic location (East, West, North, South). To handle this situation, each group except one needs a

[20] While high correlations among independent variables can indicate collinearity, it can also miss it. It is possible to have collinearity without high correlations among independent variables and vice-versa. For a more thorough check for collinearity, use SPSS's Collinearity Diagnostics, available under "Statistics" in the Regression command.

dummy variable. To dummy code the variable "Race" with "White" as the excluded category, the following variables would be necessary: the variable "Black" would be coded 0/1, with all African American respondents coded as 1 and all other respondents with known ethnicity coded 0; the variable "Hispanic," would be coded 0/1 with all Hispanic respondents coded 1 and all others coded 0. We would do this for all but one racial group called the **reference group** that has the value of 0 in all of the computed dummy variables. Then all of these dummy variables would be included in the model.

In this example, we would exclude Whites as the reference group, and the regression output would allow us to compare other ethnicities to Whites.[21] For example, if we were predicting how much money people saved for retirement and the variable "Black" had a coefficient –3,200, it would mean that Blacks save, on average, $3,200 less than Whites. If the variable "Asian" had a coefficient of 6,700, it would mean that Asians save $6,700 more for retirement than Whites. In each case, the dummy variable is interpreted relative to the reference group in the regression.

Outliers

Linear regressions can be greatly influenced by **outliers**, atypical cases. Outliers can "pull" the equation away from the general pattern, and unduly sway the regression output. But what is the appropriate way to deal with an outlier?

In some situations it is reasonable to simply delete an outlier from the analysis. For example, perhaps the outlier was a mistake – a data entry or data recording error. Another example is when there are special circumstances surrounding a specific case. For example, Nevada has a very high divorce rate. Because of its laws, it attracts people from outside the state for "drive by divorces." Because Nevada's divorce rate does not accurately measure the rate of divorce for Nevada residents, as do the divorce rates in other states, it may be reasonable to completely remove Nevada from any analysis of divorce rates.

In other situations, outliers should remain in the analyses, as an outlier may be the most important observation. In these circumstances, it is a good idea to run the analysis twice, first with the outlier in the regression and second with it excluded. If the outlier is not exerting an undue influence on the outcomes, both models should reasonably coincide. If, however, the results are vastly different, it is best to include both results in the text of the report.

Causality

Finally, researchers should use great caution in interpreting the outcomes of linear regressions as establishing causal relationships. As discussed in Chapter 1, three things are necessary to establish cause: association, time order, and nonspuriousness. While regressions show associations, they usually do not document time order. Note, how we were careful to say that poverty is associated with teen births, but did not assert that it causes them (although this might be true).

---

[21] In deciding which group to omit, it is important that there be a sufficient number of cases in the data set to allow a meaningful comparison (e.g., we would not select Whites as a comparison group if there were only a few represented in the data). It is also important that the reference group be useful in its potential for meaningful comparisons. For this reason, we chose not to use "Other" as a comparison group because it does not represent a cohesive category of observations, but rather a mish-mash of racial/ethnic groups that don't fit into the larger categories present in the data set.

One of the strengths of multiple linear regressions is that researchers can include factors (if they are available) that can control for spurious effects. However, there always remains the possibility that a spurious factor remains untested. Even though multiple variables may be included in the statistical model, it is still possible to have spurious relationships if important variables are left out. Only a large body of research would be able to account for enough factors that researchers could comfortably conclude causality.

Summary

Multiple linear regression has many advantages, as researchers can examine the multiple factors that contribute to social experiences and control for the influence of spurious effects. They also allow us to create refined graphs of relationships through regression lines. These can be a straightforward and accessible way of presenting results.

Knowing how to interpret linear regression coefficients allows researchers to understand both the direction of a relationship (whether one variable is associated with an increase or a decrease in another variable) and strength (how much of a difference in the dependent variable is associated with a measured difference in the independent variable). Knowing about the R-square helps researchers understand the explanatory power of statistical models. As with other statistical measures, the significance tests in regressions address the concern of random variation and the degree to which it is a possible explanation for the observed relationships.

As regressions are complex, care is needed in performing them. Researchers need to examine the variables and construct them in forms that are amenable to this approach, such as creating dummy variables. They also need to examine findings carefully and test for concerns such as collinearity or patterns among residuals. This being said, linear regressions are quite forgiving of minor breaches of these assumptions and can produce some of the most useful information on the relationships between variables.

Key Terms

Adjusted R-Square
B
Coefficient
Collective effects
Collinearity
Constant
Constant variance
Control variables
Degrees of freedom
Dummy variables
Explanatory power
Intercept
Linear relationship
Outliers
Normality of residuals
Outliers
Reference group
Residuals
R-square
Slope
Spurious factors
Unstandardized coefficient

Chapter 7 Exercises

Name______________________________________ Date__________________

1. Using the STATES data, test the hypothesis that states with large African American populations receive lower educational funding than predominantly White states. Test this hypothesis by performing a bivariate linear regression on EDS133 (Per Capita State and Local Gvt Spending for Education 2004) and DMS441 (Percent of Population Black: 2005). Fill in the following statistics:

   Regression Coefficient B for DMS441 ____________

   Significance level ____________

   Is the relationship significant? Yes No

   R Square ____________

   In your own words, describe the relationship between EDS133 and DMS441. How would you explain these findings?

2. Using the STATES data set, examine the relationship between states' rate of U.S. Military Fatalities in Iraq as of January 2007 (DFS85) and their 2006 Public High School Graduation Rate (EDS128).

Regression Coefficient B for EDS128	______________
Significance	______________
Is the relationship significant?	Yes No
R Square	______________

In your own words, describe the relationship between DFS85 and EDS128. How would you explain these findings?

3. Perform a regression on the relationship between the property crime rate (CRS43) as predicted by the percent of the population of a state that is living below the poverty level (PVS493).

Regression Coefficient B for PVS493 ______________

Significance ______________

Is the relationship significant? Yes No

R Square ______________

In your own words, describe the relationship between CRS43 and PVS493. How would you explain these findings?

4. Perform a multiple regression to determine if social stress predicts the crime rate (CRS28). Include as indicators of potential social stress the homeownership rate (ECS424), the divorce rate (DMS483), and the personal bankruptcy rate (ECS98).

Constant ________________

Adjusted R Square ________________

Regression Coefficient B for ECS424 ________________

Significance ________________

Is the relationship significant? Yes No

Regression Coefficient B for DMS483 ________________

Significance ________________

Is the relationship significant? Yes No

Regression Coefficient B for ECS98 ________________

Significance ________________

Is the relationship significant? Yes No

In your own words, describe these relationships and whether the theory that social stress predicts the crime rate is supported by these statistics.

5. Using the output from the regression of social stress and the crime rate in Exercise 4, write the formula which would be used to generate a line showing the association of the divorce rate with the crime rate. Hold constant ECS424 and ECS98 at their mean values. You will need to generate the mean values using the *Descriptive Statistics - Descriptives* command.

$$\hat{Y} \quad = \quad A \quad + \quad B_1(X_1) \quad + \quad B_2(X_2) \quad + \quad B_3(X_3)$$

6. Compute a new variable "CRIME2" using the above equation. Label CRIME2 "Predicted Crime rate 2." Graph and print the relationship between CRIME2 and DMS483 using the command:

Graphs

Scatter

Simple

Y Axis: CRIME2

X Axis: DMS483

7. Using the output from the regression of social stress and the crime rate in Exercise 4, write the formula which would be used to generate a line showing the association of the home ownership with the crime rate. Hold constant DMS483 and ECS98 at their mean values. You will need to generate the mean values using the *Descriptive Statistics – Descriptives* command.

$$\hat{Y} \quad = \quad A \quad + \quad B_1(X_1) \quad + \quad B_2(X_2) \quad + \quad B_3(X_3)$$

8. Compute a new variable "CRIME3" using the above equation. Label CRIME3 "Predicted Crime Rate 3." Graph and print the relationship between CRIME3 and ECS424 using the command:

Graphs
　　Scatter
　　　　Simple
　　　　　　Y Axis: CRIME3
　　　　　　X Axis: ECS424

Chapter 8
Multivariate Analysis with Logistic Regression

Overview

The previous chapter explained multivariate analysis using linear regression. This chapter introduces multivariate analysis using logistic regression. Like linear regression, logistic regression models the relationship of multiple independent variables to a dependent variable. With it, researchers can graph regression lines and use the values of independent variables to predict dependent variables. But unlike linear regression, logistic regression requires a different type of dependent variable, and its coefficients have different interpretations. In this chapter, we illustrate how to perform logistic regressions, understand output, and construct graphs and predict values based on model summary statistics.

What Is Logistic Regression?

Both linear and logistic regression analyze relationships between multiple independent variables and a single dependent variable. Whereas linear regression requires a scale dependent variable with a normal distribution, **logistic regression** requires a **binary dependent variable** —a categorical variable with two categories. Like dummy variables, these are coded 0/1 and indicate if a condition is or is not present, or if an event did or did not occur. Mortality, for example, would be coded 0 (alive) and 1 (dead); recidivism (repeat crimes) would be coded 0 (no new crimes) and 1 (committed a new crime).

Logistic regression uses the independent variables to estimate the likelihood of occurrence of one of the categories of the dependent variable. For example, a researcher can use information about criminals, such as the type of crime they committed, the amount of time they spent in jail, and their age, to model the likelihood of a repeat offense occurring. Although no model can predict the behavior of any individual criminal, it gives the likelihood that a certain type of person will become a repeat offender. For instance, what is the likelihood of a young female nonviolent criminal becoming a recidivist compared to an older male violent criminal? This is a question that logistic regression can test.

Logistic regression has the same advantages as linear regression, including the ability to construct multivariate models and include control variables. It can use both scale and dummy variables as predictors. But the chief differences with linear regression are that logistic regressions use binary dependent variables and that they predict likelihoods. Recall that the coefficient in a linear regression indicates the difference in the dependent variable associated with one unit difference in the independent variable. In contrast, output in a logistic regression help us understand the chances that an event will or will not occur, and the impact that a one-unit change in an independent variable will have on the likelihood of that event occurring.

When Can I Use a Logistic Regression?

Perhaps the most important consideration in the use of logistic regression is the structure of the dependent variable. Logistic regressions require binary dependent variables. Binary variables indicate if a condition is present or absent. Consider this fact: many types of events do not fall on continua; they exist in binary form. People are either living or dead, they have children or they do not, students either graduate or they do not, people use drugs or they do not, states have the death penalty or they do not. Predicting outcomes—the likelihood that observations fall into one condition or the other—is an ideal use of logistic regression. As such, this analytic tool is among the most valuable in medical research, which often seeks to establish if a treatment has proven itself to be effective.

Logistic regression can also be useful when the dependent variable has more than two categories. For example, a variable that measures the highest degree earned may have values 1 "High School" 2 "Bachelors" 3 "Masters" 4 "Ph.D." A researcher could recode this variable to a binary form: 0 "High School" or 1 "Bachelors or higher" to predict the likelihood of earning at least a college degree.

Even some scale variables can be recoded so that they take on binary forms. One way of categorizing scale variables is to indicate if each observation has a high value or a low value. A study by Straus and Sweet (1992), for example, studied the incidence of verbal aggression — yelling and name-calling — in American families. Most couples reported never yelling at their partner, but some couples yell and threaten every day. The dependent variable (number of verbal assaults in a year) was very skewed, making linear regression inappropriate. Their solution was to recode the scale variable into a binary categorical variable. Verbally aggressive couples, which yelled or threatened more than once per month (13+ per year), were coded as "1." The comparatively non-aggressive couples were coded as "0." Straus and Sweet used logistic regression to examine the influence of gender, socioeconomic status, children, drinking, and drugs on the likelihood of high levels of verbal aggression occurring, and revealed relationships that were otherwise unrecognizable in linear regression output.

When constructing binary variables, an important concern is that the categories are **mutually exclusive**—a case cannot be in more than one category at the same time. For example, a variable that indicates whether a person owns a truck or a car cannot be coded as 0 "truck" and 1 "car," because a person could own both (thereby falling into both categories). A better approach would be to create two binary variables. One would indicate if the person owned a car, coded 0 "does not own" and 1 "owns." The second would indicate if the person owned a truck, coded 0 "does not own" and 1 "owns." Two logistic regressions could predict car ownership and truck ownership.

Understanding Relationships through Probabilities

Logistic regressions predict likelihoods, which can be measured by **probability**, **odds**, or **log-odds**. In conversation, people use "probabilities" and "odds" interchangeably, but there is an important distinction. A probability is the ratio of the number of occurrences to the total number of possibilities. In contrast, the odds is the ratio of the number of occurrences to the number of non-occurrences. For example, what is the likelihood of drawing a spade from a deck of cards? There are fifty two cards in a deck (52 possibilities). Thirteen of those cards are spades (13 occurrences) and 39 cards are hearts, diamonds, and clubs (39 non-occurrences). The *probability* of choosing a spade is 13/52, or 1/4. One fourth of the cards are spades. The *odds* of choosing a spade is 13/39 or 1/3. For every spade in the deck, there are three non-spades.

As another example, the probability of catching the flu during any given winter is about 0.10, or 1/10. This means that out of every ten people, approximately one will catch the flu. In contrast, the odds of catching the flu during any given winter is 1/9—for every one person who catches the flu, nine do not. It is easy to convert back and forth between probability and odds, as they give the same information. Table 8.1 shows the correspondence between some probabilities and odds.

Number of occurrences in 100 possibilities	Probability	Odds
0	0	0
10	1/10	1/9
25	1/4	1/3
33	1/3	1/2
50	1/2	1
75	3/4	3
90	9/10	9
100	1	∞ (infinity)

Table 8.1 Common Values of Probability and Odds

As you can see, probabilities range from 0 to 1, whereas odds range from 0 to infinity. An odds of one means equal probability of occurrence and non-occurrence (.50). An odds less than 1 means occurrence is less likely than a non-occurrence. An odds greater than 1 means occurrence is more likely than non-occurrence. Distinguishing probabilities from odds is important for both accuracy in reporting findings and interpreting coefficients and graphs. Even when findings are reported as odds, the following formula converts them to probabilities:

$$Probability = \frac{Odds}{1 + Odds}$$

We will use this formula to convert the logistic regression output to predicted probabilities.

Unfortunately, if we set up the model to predict the probability of occurrence directly, the coefficients would be too complicated to calculate or interpret. Luckily, there is a function of the probability that has a nice, linear relationship with the independent variables. If we use this function of probability, the model has the exact same form as a linear regression model, and the coefficients are simple to calculate. This function is the **logit function** (hence, the name logistic regression), also called the **log-odds** function.

Linear Regression	Logistic Regression
Y = A+B(X)	log-odds = A+B(X)

The log-odds is the natural logarithm of the odds of occurrence. The disadvantage is that the coefficients are on the log scale and a little trickier to interpret. Converting them back to their odds form makes it a bit easier. As we illustrate the application of logistic regression below, you will see why the log-odds are reported, and how to convert them back to probability scores.

Logistic Regression: A Bivariate Example

To illustrate logistic regression, we will use the GSS04 data. Our interests are in predicting the types of people most likely to report owning a gun. For this bivariate example, we examine the influence of political views. The guiding hypothesis is that conservative people are more likely to own guns because political conservatives tend to support gun ownership rights.

The first step in the analysis is to check that the dependent and independent variables have structures that are suitable for logistic regression. The dependent variable OWNGUN should be a 0/1 binary variable, with 1 indicating gun ownership. To check this, perform:

Analyze
    Descriptive Statistics
        Frequencies
            Variables: OWNGUN
            OK

It confirms that OWNGUN has two values: No and Yes, coded 0 and 1, respectively. The remaining values are coded as missing.

The independent variables in a bivariate logistic regression can be scale or dummy coded. The *Frequencies* command checks this as well. The variable POLVIEWS is an ordinal variable, locating respondents on a scale ranging from 1 (extremely liberal) to 7 (extremely conservative). Higher values indicate the respondent is more conservative. Although ordinal variables are actually categorical, since there are many values that measure an underlying continuum, it is reasonable to treat it as a scale variable. We expect a positive relationship between POLVIEWS and OWNGUN. We can now test the relationship:

Analyze
    Regression
        Binary Logistic
            Dependent: OWNGUN
            Covariates: POLVIEWS
            OK

Within your output should be three boxes that look like Figure 8.1. We focus on the odds ratios, coefficients, and significance tests.

Dependent Variable Encoding

Original Value	Internal Value
0 NO	0
1 YES	1

Omnibus Tests of Model Coefficients

		Chi-square	df	Sig.
Step 1	Step	47.793	1	.000
	Block	47.793	1	.000
	Model	47.793	1	.000

Variables in the Equation

		B	S.E.	Wald	df	Sig.	Exp(B)
Step 1[a]	POLVIEWS	.362	.055	44.097	1	.000	1.436
	Constant	-2.088	.249	70.573	1	.000	.124

a. Variable(s) entered on step 1: POLVIEWS.

Figure 8.1 Logistic Regression Output

Interpreting Odds Ratios and Logistic Regression Coefficients

The first step is to make sure that SPSS did what we thought it did. The *Dependent Variable Encoding* table shows the values of the dependent variable. The "Original Value" column lists the values as they appear in the data set and the "Internal Value" column lists the values that SPSS assigned to them for the logistic regression. SPSS always assigns internal values of 0 and 1 and the output is always in terms of the likelihood of an individual having the value of "1." In this example, SPSS has assigned the same values that we had, so we know that it is predicting the likelihood of owning a gun.

We now consider the interpretation of the **logistic regression coefficients** or **log-odds**. Logistic regression coefficients, in the column "B" in the *Variables in the Equation* box, have the same purpose as linear regression coefficients—they show the direction and strength of the relationship between the independent and dependent variables. The direction in this example is positive (.362), indicating that the more conservative a person is, the greater the likelihood of owning a gun. The strength of the relationship is hard to gauge with the coefficients, since they are measured on the log scale. They represent the difference in the log-odds of the dependent variable for each one-unit change in the independent variable. In other words, for every movement from one increment to the next (toward the direction of increasing conservativeness) on the political orientation scale, the log-odds of gun ownership increase by .362.

As most people (including us!) can't think on a log scale, a more intuitive measurement of the strength is the odds ratio, shown in the column Exp(B) in the *Variables in the Equation* box. The **odds ratio** is the coefficient with the log removed. Because of the mathematical properties of logs, it measures *how many times* higher the odds of occurrence are for each one-unit increase in the independent variable. The odds ratio for POLVIEWS is 1.436. Since it is greater than 1, a higher score (being more conservative) increases the odds of owning a gun. How much more? Each one-unit increase on the political views scale increases the odds of owning a gun by a factor of 1.436. People who consider themselves liberal (with a score of 2) are 1.436 times as likely to own a gun as people who consider themselves very liberal (with a score of 1). Likewise, people who consider themselves slightly liberal (with a score of 3) are 1.436 times as likely to own a gun as people who consider themselves liberal (with a score of 2), and so on.

The next step is to see if the relationship is statistically significant. The *Sig.* column shows a significance level for POLVIEWS of .000. The probability that the observed relationship can be attributed to chance is extremely low, and we conclude there is a relationship between political views and gun ownership.

Using Logistic Regression Coefficients to Make Predictions

Like linear regressions, logistic regression coefficients can be used to make predictions. Suppose that we wanted to know the probability of gun ownership for someone who is extremely conservative (POLVIEWS=7). To do this requires plugging in a value of seven for POLVIEWS into the logistic regression equation and converting the log-odds to probabilities.

The logistic regression equation is:

$$\textit{Log-odds} = A + B(X) \qquad (1)$$

Because the logistic regression represents the log-odds, we have to take the antilog (Exp) as we convert the formula to represent probability, as shown in this equation:

$$\textit{Odds} = \textit{Exp}(A+B(X)) \qquad (2)$$

Recall that the formula for converting odds to probabilities is:

$$\textit{Probability} = \frac{\textit{Odds}}{1+\textit{Odds}} \qquad (3)$$

Putting it all together by replacing *Odds* with equation 2, gives:

$$\textit{Probability} = \frac{\textit{Exp}(A+B(X))}{1+\textit{Exp}(A+B(X))} \qquad (4)$$

In this example:

$$\textit{Probability} = \frac{\textit{Exp}(-2.088+.362(7))}{1+\textit{Exp}(-2.088+.362(7))}$$

$$\textit{Probability} = \frac{\textit{Exp}(.446)}{1+\textit{Exp}(.446)}$$

$$\textit{Probability} = \frac{1.562}{2.562}$$

Probability of a very conservative person owning a gun = .61

Likewise, we could predict gun ownership for someone who is extremely liberal (POLVIEWS=1):

$$Probability = \frac{Exp(A + B(X))}{1 + Exp(A + B(X))}$$

$$Probability = \frac{Exp(-2.088 + .362(1))}{1 + Exp(-2.088 + .362(1))}$$

$$Probability = \frac{Exp(-1.726)}{1 + Exp(-1.726)}$$

$$Probability = \frac{.178}{1.178}$$

$$Probability\ of\ a\ very\ liberal\ person\ owning\ a\ gun = .15$$

Why take the trouble to make these calculations? One reason is to make findings meaningful to non-statisticians and to help general audiences understand the meaning of logistic regression results. We can succinctly state our findings in the following way.

> *Gun ownership is significantly associated with political attitudes. The probability of a very conservative person owning a gun is .61. The probability of a very liberal person owning a gun is much lower, .15.*

Using Coefficients to Graph a Logistic Regression Line

Another way of making results accessible (and to interpret their meaning) is to graph them. The approach to graphing logistic regression results, while it may appear complicated on the surface, is actually similar to the approach we used for linear regression. The only difference is that the calculations accommodate the conversion of log-odds into probabilities. To graph the logistic regression, we will find the predicted probabilities for different values of the independent variable. Unlike the linear relationship in linear regressions, the logistic regression graphs will show a **sigmoidal** or **S-shaped** relationship, as represented in Figure 8.2.

The reason that probabilities are sigmoidal, not linear, can be illustrated by the relationship between the probability of injury in a car accident and driving speed. Think of the X-axis in Figure 8.2 as speed in miles per hour (mph), and the Y-axis as the probability of being seriously injured. People traveling at *any* low speed are unlikely to be seriously injured. The probability of serious injury changes little between one mph and six mph. Likewise, people traveling at *any* high speed are very likely to be seriously injured, so probability changes little at this end of the graph, as it is a near certainty that injuries will happen.

Now consider a 5 mph change in the middle of the graph. Traveling 65 versus 60 mph greatly affects the probability of injury in a crash. Notice that for probabilities between .2 and .8, the curve is pretty much linear, but at the ends, it levels off. Pay attention to the range of probabilities as you look at the graphs in this chapter.

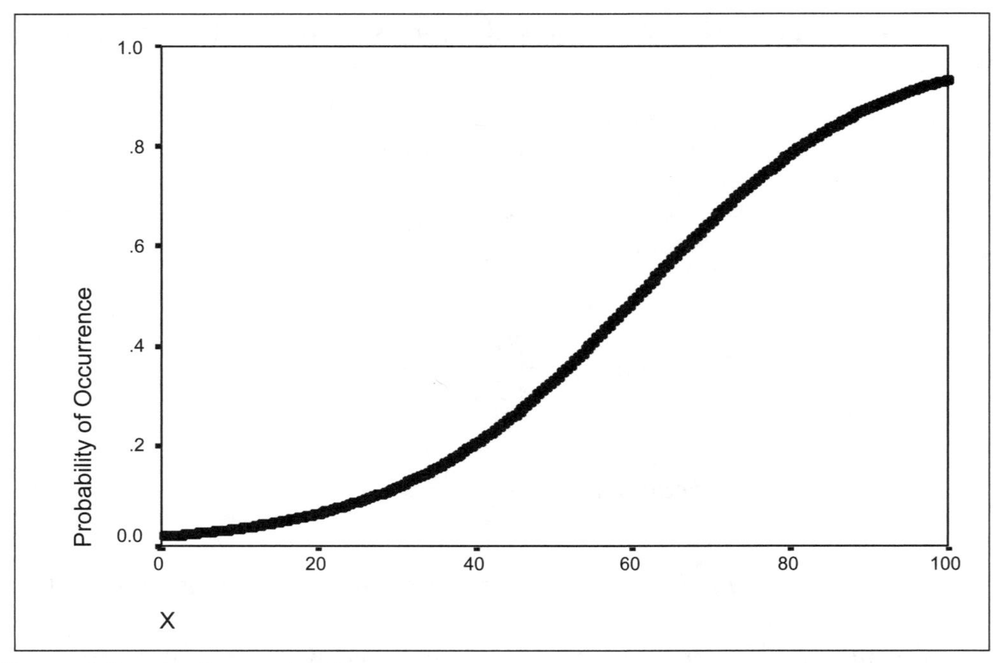

Figure 8.2 A Sigmoidal Curve

To make the logistic regression graphs, we use the same formula we used earlier to find predicted probabilities. Rather than compute the predicted probabilities by hand, we will allow SPSS to calculate the predicted probabilities from the constant and coefficient using the *Compute* command. Be careful to place parentheses "( )"correctly in the formula, as mistakes will produce errors.

Transform
Compute Variable
Target Variable: GUNPRE
Numeric Expression: (*Exp* (-2.088 + (.362*POLVIEWS)))/ (1+*Exp* (-2.088 + (.362*POLVIEWS))).
Type&Label
Label: Predicted Gun Ownership
Continue
OK

Then we use GUNPRE to graph the logistic regression line, using the *Scatterplot* commands (Figure 8.3).

Graphs
Interactive
Scatterplot

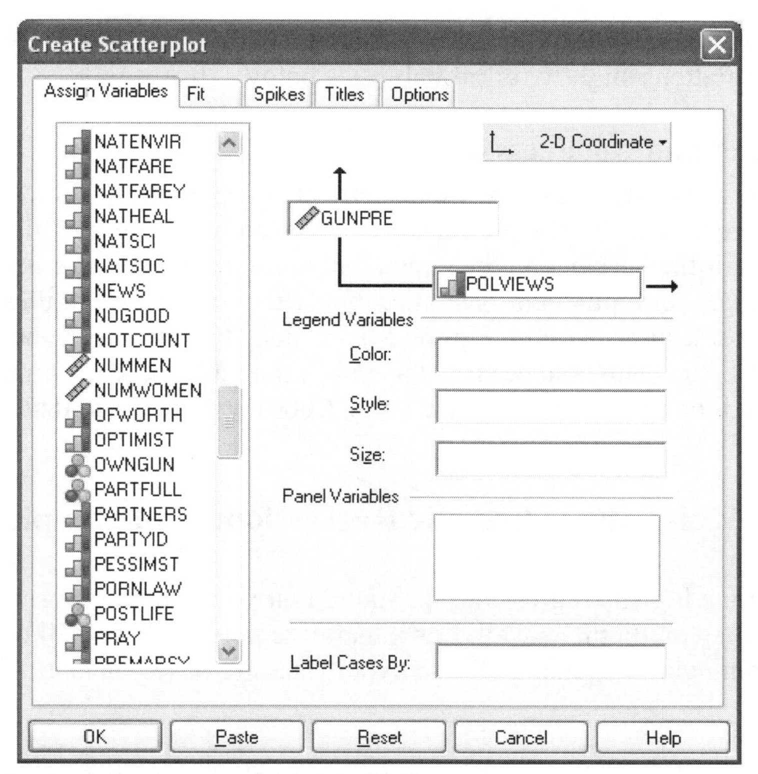

Figure 8.3 Scatterplot Commands

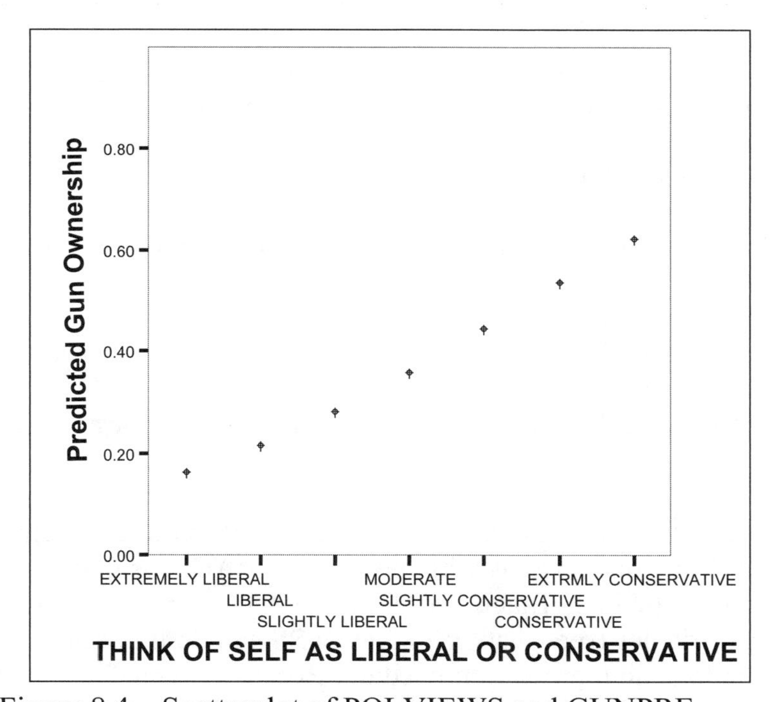

Figure 8.4 Scatterplot of POLVIEWS and GUNPRE

Figure 8.4 shows the degree to which political attitudes are associated with gun ownership. In this example, the graph represents the central portion of the sigmoid curve, as all political groups have between a .15 and a .65 probability of owning a gun. Notice how the curve is mostly linear, but flattens slightly for probabilities below .20.

Model Chi-Squares and Goodness of Fit

There is no statistic in logistic regression that performs the function of the R-square statistic of linear regression. Recall that the R-square statistic measures goodness of fit and tells how much the model accounts for the variance in the dependent variable. There have been some attempts by statisticians to create an equivalent statistic, but each has shortcomings. We recommend examining the Model Chi-Square in Figure 8.1, which tests whether the model as a whole predicts occurrence better than chance. It is located in the table called "Omnibus Tests of Model Coefficients." Since it has a low significance value (.000), we conclude that the model does have predictive power.

Multivariate Logistic Regression: An Example

In the bivariate logistic regression, we found that political views predict gun ownership, but other variables may predict it as well. For illustrative purposes, we will offer the following hypotheses concerning gun ownership. These hypotheses are based on stereotypes of different groups in society.

Hypothesis 1: Political conservatives are more likely to own guns than liberals

Hypothesis 2: African Americans are more likely to own guns than White Americans

Hypothesis 3: Gun ownership is least prevalent among those highest on the socioeconomic scale, and most common among those at the lower levels.

We will use the following variables:

POLVIEWS (A scale variable with high values indicating more conservative orientations)

RACE (Race of respondent – a categorical variable)

SEI (Socioeconomic Index – a scale variable with higher values indicating higher social status)

Once again, the first step is to check the variable structures using the *Frequencies* command. The results bring up concerns with the variable RACE, as that is a categorical variable that has three categories (White, Black and Other). One approach would be to recode this variable into dummy variables (see Chapter 7), but SPSS can do this automatically if we specify it as categorical in the logistic regression procedure. Because "White" is the first category in the variable RACE, and because we want to gauge the likelihood of gun ownership of African Americans relative to that of White respondents, we select "first" as the reference category in the following commands that run the multivariate logistic regression. Components of your output should look like Figure 8.5.

Analyze
Regression
Binary Logistic
Dependent: OWNGUN
Covariates: POLVIEWS
RACE
SEI
Categorical:
Categorical Covariates: RACE
Reference Category: First, Change (while RACE is highlighted)
Continue
OK

Case Processing Summary

Unweighted Cases [a]		N	Percent
Selected Cases	Included in Analysis	864	30.7
	Missing Cases	1948	69.3
	Total	2812	100.0
Unselected Cases		0	.0
Total		2812	100.0

a. If weight is in effect, see classification table for the total number of cases.

Dependent Variable Encoding

Original Value	Internal Value
0 NO	0
1 YES	1

Categorical Variables Codings

			Parameter coding	
		Frequency	(1)	(2)
RACE RACE OF RESPONDENT	1 WHITE	691	.000	.000
	2 BLACK	112	1.000	.000
	3 OTHER	61	.000	1.000

Variables in the Equation

		B	S.E.	Wald	df	Sig.	Exp(B)
Step 1 [a]	POLVIEWS	.353	.055	41.090	1	.000	1.423
	SEI	.010	.003	9.787	1	.002	1.010
	RACE			19.838	2	.000	
	RACE(1)	-.845	.250	11.402	1	.001	.430
	RACE(2)	-1.095	.348	9.882	1	.002	.335
	Constant	-2.388	.306	60.860	1	.000	.092

a Variable(s) entered on step 1: POLVIEWS, SEI, RACE.

Figure 8.5 Multivariate Logistic Regression Output

Interpreting Logistic Regression Output

Once again, the first step is to make sure that SPSS did what we wanted. The *Dependent Variable Encoding* table (Figure 8.5) shows that OWNGUN is coded correctly — "YES" has an internal value of 1. Also, we look to see that sufficient cases have been included in the analysis. The *Case Processing Summary* table shows that there are many missing cases, but most of these can be attributed to missing values in the original dependent variable, and are not the result from the unique combination of variables included in the model.

The *Categorical Variables Codings* table has the information for the categorical independent variable RACE. The "Parameter Coding" columns refer to the dummy variables. Since RACE has three categories, it has two dummy variables, and entries in two columns. The *Variables in the Equation* table lists two variables under RACE: RACE(1) and RACE(2); each refers to one of the dummy variables. RACE(1) has values of 1 for Blacks and 0 for everyone else. RACE(2) has values of 1 for Others and 0 for everyone else. Since Whites have a value of 0 for both RACE(1) and RACE(2), it is the reference category. Therefore, in the *Variables in the Equation* table, RACE(1) refers to the likelihood of African Americans owning guns compared to Whites, and RACE(2) refers to the likelihood of "other races" owning guns compared to Whites.

All of the variables are statistically significant (see the *Variables in the Equation* table in Figure 8.5). As in the bivariate logistic regression, more conservative political views (higher values of POLVIEWS) are associated with higher likelihoods of gun ownership. The odds of gun ownership are 1.423 times higher for each one-unit increase on the POLVIEWS scale, as the Exp(B) column of the *Variables in the Equation* box shows. The odds ratio has changed only slightly from the bivariate logistic regression, even after accounting for the influence of socioeconomic status and race.

SEI (socioeconomic index) is significant at a .001 level with an odds ratio of 1.010. People who are higher on the socioeconomic index, contrary to our hypothesis, have a higher likelihood of owing guns than those who are lower on the scale.

RACE is also significant. The Sig. value for RACE tells us that the odds of gun ownership for the three race categories are significantly different, but not which ones. RACE(1) compares African Americans and Whites, RACE(2) compares Others to Whites. The odds ratios reveal that African Americans are .430 times (about half) as likely to own a gun as Whites and other minorities are .335 (about 1/3) as likely to own a gun as Whites. Again, our hypothesis that racial minorities are more likely to own guns was not only unsupported, an opposite conclusion is reached. African Americans and other racial minorities are considerably less likely to own guns than Whites.

Using Multivariate Logistic Regression Coefficients to Make Predictions

Logistic regression coefficients, like linear regression coefficients, can be used to make predictions. To predict the likelihood that a very liberal White person owns a gun, we insert into the regression equation the mean value of SEI, 0 for RACE1, 0 for RACE2, 1 for POLVIEW, and the logistic regression coefficients to calculate the log-odds. We can then convert the log-odds to a probability.

The numbers for the equations are as follows:

Constant = -2.388

	B	Value	
POLVIEW	.353	4.23	(mean value)
RACE1	- .845	0	(value of White)
RACE2	-1.095	0	(value of White)
SEI	.010	47.86	(mean value)

log-odds = A + B1(X1) + B2(X2) + B3(X3) + B4(X4)

log-odds = A + B1(POLVIEW) + B2(RACE1) + B3(RACE2) + B4(SEI)

log-odds = -2.388 + .353(POLVIEW) - .845(RACE1) - 1.095(RACE2) + .01(SEI)

log-odds = -2.388 + .353(1) - .845(0) - 1.095(0) + .01(47.86)

log-odds = -1.56

odds = exp(log-odds) = Exp(–1.56)

$$probability = \frac{Exp(-1.56)}{1 + Exp(-1.56)}$$

$$probability = \frac{.21}{1.21} = .17$$

We conclude that a very liberal White person with average socioeconomic status has a .17 probability of owning a gun.

Using Multivariate Coefficients to Graph a Logistic Regression Line

Like bivariate logistic regression lines, multiple logistic regression lines use the formula we just used to predict probabilities. In this case, we will create two regression lines predicting the probabilities of gun ownership by socioeconomic status, one line for White respondents and the other line for African American respondents. We hold the other variables constant at selected values, but allow SEI to vary instead of inserting its mean.

Log-odds = -2.388 + .353(POLVIEW) + .010(SEI) - .845(RACE1) - 1.095(RACE2)

To create these two regression lines, we use the *Compute* procedure, and include in the formula the conversion of the log-odds back to probabilities:

White respondents:

Transform

Compute Variable
Target Variable: GUNPRE2
Numeric Expression: (Exp (-2.388 +(.353*4.23) + (.010*SEI) +(- .845*0) + (-1.095*0)))/(1+Exp (-2.388 + (.353*4.23) + (.010*SEI) + (- .845*0)+(- 1.095*0)))
Type and Label:
Label: Predicted Gun Ownership by Socioeconomic Status White
Continue
OK

African American respondents:

Transform

Compute Variable
Target Variable: GUNPRE3
Numeric Expression: (Exp (-2.388 +(.353*4.23) + (.010*SEI) +(- .845*1) + (-1.095*0)))/(1+Exp (-2.388 + (.353*4.23) + (.010*SEI) + (- .845*1)+(- 1.095*0)))
Type and Label:
Label: Predicted Gun Ownership by Socioeconomic Status Black
Continue
OK

The results can be graphed with a *Scatter/Dot* in the Chart Builder, creating an overlay scatterplot (Figure 8.6).

Graphs
Chart Builder

Your results should look like much like Figure 8.7 after adding titles and other formatting refinements. Notice in this graph that the slopes of the lines for Whites and African Americans are different. This is due to their sigmoidal construction of probability estimates. This type of visual representations offers compelling presentations of the data and can be valuable contributions to reports.

As you can see, graphing multivariate lines can be a labor intensive process. It requires setting up the equation, transferring coefficients from output, and converting the log-odds into probabilities correctly. There are numerous opportunities for error. A few tricks can check for mistakes. First, the values on the Y-axis must be between 0 and 1, the minimum and maximum values for probabilities. If they exceed this range, the regression line was incorrectly computed. Second, the direction of the relationship should match the direction indicated in the regression coefficients. SEI's positive coefficient means that the line should flow from the lower left corner of the graph to the upper right.

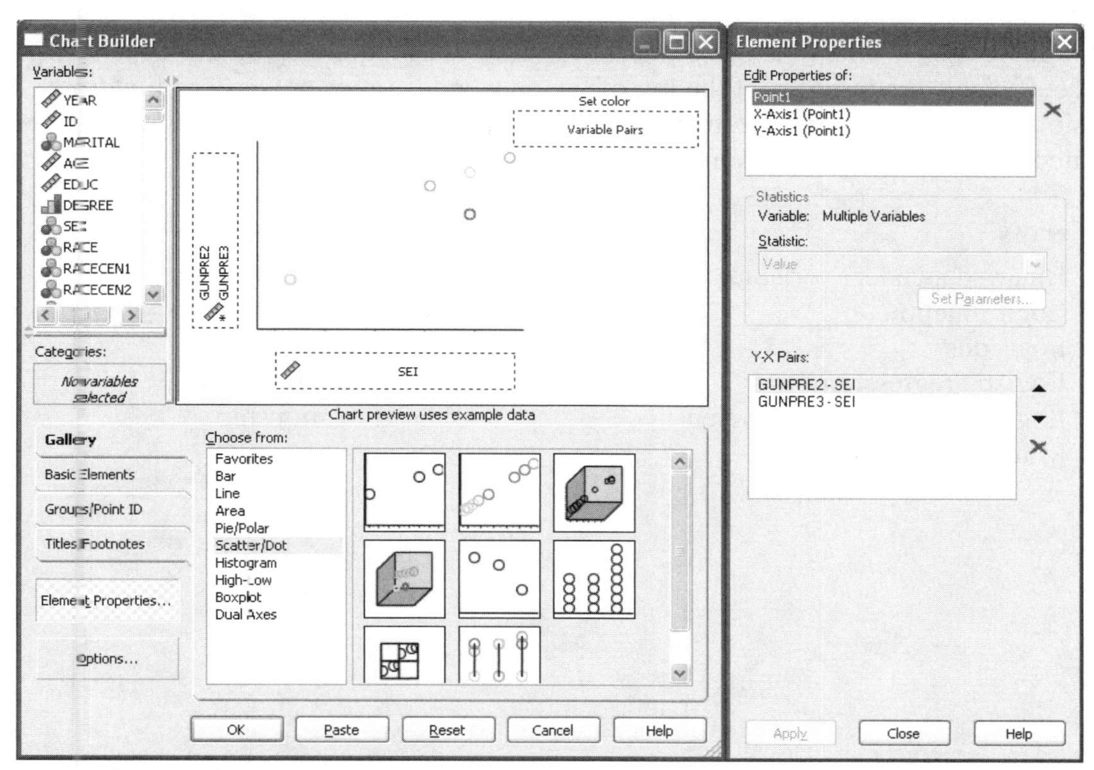

Figure 8.6 Chart Builder Commands to Create an Overlay Scatterplot of Constructed Variables

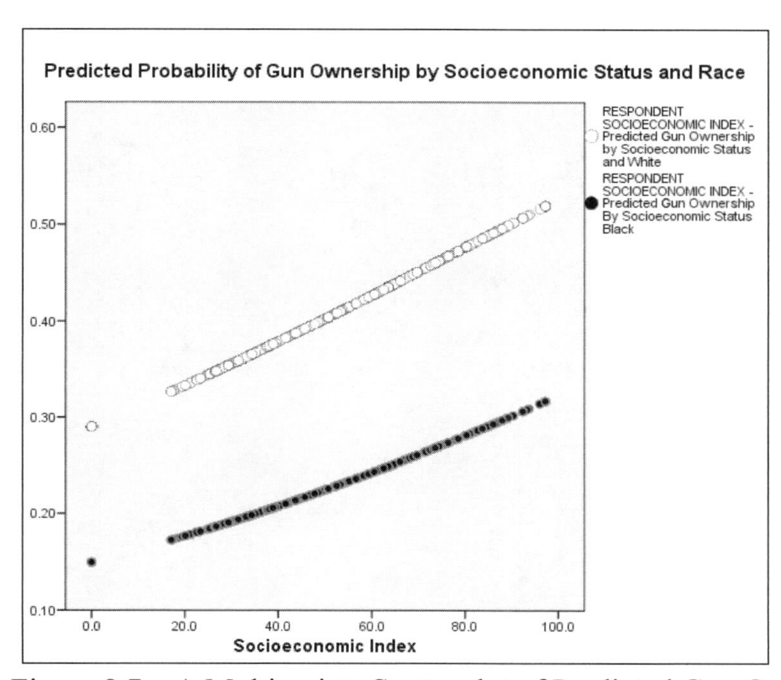

Figure 8.7 A Multivariate Scatterplot of Predicted Gun Ownership

Summary

Logistic regression works much like linear regression but requires a binary dependent variable. Since coefficients are in terms of the log of the odds, odds ratios are easier to interpret. Logistic regressions can be graphed by converting the log-odds, and predictions of probabilities can be calculated for specific sets of conditions.

Key Terms

- Binary dependent variable
- Logit function
- Log-odds
- Logistic regression
- Logistic regression coefficients
- Mutually exclusive
- Odds
- Odds ratio
- Probability
- Sigmoidal relationship
- S-shaped relationship

Chapter 8 Exercises

Name____________________ Date ________________

1. Using the GSS04 data and the variable GRASS, examine whether the respondent's socioeconomic index score (SEI) has an effect on the likelihood of supporting the legalization of marijuana. Use logistic regression to generate the following information:

   Constant ________________

   Logistic Regression Coefficient for SEI ________________

   Odds ratio for SEI ________________

   Significance ________________

   Is this relationship statistically significant? Yes No

   Do people scoring high on SEI have a greater or lesser probability of supporting legalization in comparison to people scoring low on SEI?

   Greater Lesser No Effect

2. Using the GSS04 data and the variable GRASS, examine whether the respondent's education (EDUC) has an effect on the likelihood of supporting the legalization of marijuana. Use logistic regression to generate the following information:

   Constant ________________

   Logistic Regression Coefficient for EDUC ________________

   Odds ratio for EDUC ________________

   Significance ________________

   Is this relationship statistically significant? Yes No

   Do people with high levels of education have a greater or lesser probability of supporting the legalization of marijuana in comparison to people with low education?

   Greater Lesser No Effect

3. Using the output from the previous logistic regression, predict the probability of a person being in support of legalization of marijuana if the respondent has a high school education (12 years).

   Step 1. Generate the log-odds using the formula: $Log\text{-}odds = A+B(X)$

   Log-odds _______________

   Step 2. Generate the odds using the formula: $Odds = Exp(A+B(X))$

   Odds _______________

   Step 3. Generate the probability using the formula: $Probability = \dfrac{Odds}{1+Odds}$

   Probability ____________

4. Using the GSS04 data to determine if RACE, SEI, and AGE predict attitudes toward spanking children with the variable SPANKING (Favor Spanking to Discipline a Child).

   Step 1: Recode SPANKING into a new binary variable SPANKING2, so that "Agree" and "Strongly Agree" are recoded to 1, and "Disagree" and "Strongly Disagree" are coded to 0. Report the percentages below

   0 ______________
   1 ______________

   Step 2: Run a binary logistic regression on SPANKING2, with the variables RACE, SEI, and AGE. Specify RACE as a categorical variable, with "White" as being the contrast indicator variable

   Constant ________________

   Logistic Regression Coefficient for RACE(1) ________________

   Odds ratio for RACE(1) ________________

   Logistic Regression Coefficient for RACE(2) ________________

   Odds ratio for RACE(2) ________________

   Logistic Regression Coefficient for SEI ________________

   Odds ratio for SEI ________________

   Logistic Regression Coefficient for AGE ________________

   Odds ratio for AGE ________________

   Which Relationships are Significant? ________________

Are African Americans more supportive or less supportive of spanking than Whites?

More Less No Effect

Are older people more supportive or less supportive of spanking than younger people?

More Less No Effect

Are people with higher socioeconomic statuses more supportive or less supportive of spanking than those in lower statuses?

More Less No Effect

5. Using the output from the previous logistic regression, predict the probability of a person being in support of spanking if the respondent has the average SEI (47.86), is White, and is 30 years old.

Step 1. Generate the log-odds using the formula:

$$Log\text{-}odds = A + B1(X1) + B2(X2) + B3(X3) + B4(X4)$$

Log-odds ______________

Step 2. Generate the odds using the formula:

$$Odds = Exp\ A + B1(X1) + B2(X2) + B3(X3) + B4(X4)$$

Odds ______________

Step 3. Generate the probability using the formula: $Probability = \frac{Odds}{1+Odds}$

Probability ___________

6. Using the output from the previous logistic regression, predict the probability of a person being in support of spanking if the respondent has the average SEI (47.86), is African American, and is 30 years old.

Step 1. Generate the log-odds using the formula:

$$Log\text{-}odds = A + B1(X1) + B2(X2) + B3(X3) + B4(X4)$$

Log-odds ______________

Step 2. Generate the odds using the formula:

$$Odds = Exp\ A + B1(X1) + B2(X2) + B3(X3) + B4(X4)$$

Odds ______________

Step 3. Generate the probability using the formula: $Probability = \frac{Odds}{1 + Odds}$

Probability ____________

7. Generate and print a scatterplot showing the predicted relationship between SPANKING2, SEI, AGE, and RACE. Place SEI on the X axis, specify AGE to be 30 years, and generate two lines, one predicting the probabilities for African Americans and the other for Whites.

   Do your answers in questions 5 & 6 correspond with this graph? Yes No

Chapter 9
Writing a Research Report

Overview

After completing data analysis, researchers conclude their projects by writing **research reports**. These reports are organized presentations of all phases of the research projects, from beginning to end. They should present research questions, relevant literature, key findings, and the limitations of the study. Writing a good report is one of the most important phases of the research project because it is the means to make study findings known and accessible to the wider scientific community and general public.

Writing the research report can be very challenging because it requires demonstrating a mastery of the research question, the analytic methods, and the findings. But compounding this challenge is the task of explaining the analysis to an audience who may, or may not, have statistical expertise. Readers should be able to understand the complexities of the data analyses without becoming lost in the minutiae of the data analysis process. They need to understand the limitations, but also to appreciate the insights the study offers. Furthermore, many readers will not be conversant in all of the statistical methods used, and therefore analyses need to be accompanied with a written explanation of the findings. At the report phase of the project, the data analyst shifts from being an explorer to being a writer.

In this chapter, we offer an overview of the composition of a good report. The report should include a scholarly but accessible writing style and a structure that offers an organized and coherent presentation of information and interpretation of results.

Writing Style and Audience

The main purpose of a research report is to condense the questions and findings so that the reader can understand the study's implications and limitations. It should introduce the reader to the research question and offer a concise overview of the project. The first question a researcher should consider before writing the report is "who is the audience?" Not all readers have the same skills or abilities to comprehend statistics, and therefore the style needs to be appropriate for the end user. Scientists writing to other scientists can take for granted that their readers will understand fundamental statistical methods, such as those discussed in this book. However, even this audience will need to have findings and methods summarized so that they can critically evaluate the research project.

On the other hand, writing to a general audience requires an entirely different style. Statistical results will be meaningless to those who have little understanding of "coefficients," "standard errors," or "significance tests." Writing a graded research report for a college course requires yet a different style. The professors who grade these papers are concerned with not only the most important findings, but also an in-depth discussion of the process used to produce those findings. In comparison, scientists and general audiences are more interested in the findings than in the process.

In all cases, constructing a report does not consist of churning out lots of statistical output and leaving it for readers to decipher (frequently they will lack these skills). Rather, it requires analyzing lots of statistical output, weeding out analyses that are peripheral to the research question, and keeping only analyses that are central. Further, it involves reconfiguring statistics into a reader friendly format , with tables, graphs, and written text. Statistical output, such as that produced by SPSS, is designed to give researchers information for purposes beyond hypothesis testing. For instance, some statistics are used for diagnostic purposes, and as such, have little interest to readers. Therefore, researchers often need to distill hundreds of pages of output into a modest number of graphs and tables.

The Structure of a Report

Reports vary in their structures, but usually include the sections that constitute journal articles. Researchers can modify this structure depending on the needs of the audience, but even then, the following information is important:

Title
Abstract
Introduction
Literature review
Methods
Findings
Conclusion
References

Journal articles require researchers to present this information succinctly, in about twenty pages. **Evaluation reports**, which are written for practitioners or administrators, are often longer, have slightly altered section headings, but hold the same overall format. Theses and dissertations also have the same format, but the literature review, methods, and conclusion each comprise full chapters. Findings can be many chapters, each examining part of the research question. Below, we examine each of the eight sections of a research report and highlight the concerns researchers should consider as they publish their studies.

The Title

The **title** is a concise description of the research project. Because computer searches use titles to locate research articles, a title is a primary way to attract readers in the first place. A title that lacks necessary information or presents the subject unsatisfactorily may discourage potential readers. The minimal standard for a title is that it be short (5-10 words) and that it offer some information about the research topic.

It is not in a researcher's interest to pick the types of titles that appear in *Cosmopolitan* or *People* magazines. Consider the merits of these titles:

"Who Will Save the Children?"

"A Struggle Against All Odds"

"Desperate, But Not Defeated!"

While these titles are catchy, they share two problems. The first problem is that they do not describe the issue being addressed in the report. The reader does not know what is threatening children, what is the struggle against the odds, or who is desperate but not defeated. It is better to sacrifice "pizzazz" in favor of information. The second problem is that these titles imply that the researcher is an advocate of a political cause, rather than an objective analyst of data. Because our reports are generally written to the scientific community, or to people who trust our attempts at objectivity, it is to the writers' advantage to write in a non-inflammatory style.

Here are a few more titles, drawn from the American Sociological Review:

"Cultural Context, Sexual Behavior, and Romantic Relationships in Disadvantaged Neighborhoods"

Social Boundaries and Marital Assimilation: Interpreting Trends in Racial and Ethnic Intermarriage

Women's Political Representation, 1893-2003

These titles are less "catchy," but they are much better than the previous titles. Each title offers substantial information about the article's research topic. Furthermore, the tone implies that the study will offer a balanced evaluation of the data.

The Abstract

Although the abstract is the first major section of a research report, it is usually written last. This is because the **abstract** is a concise summary of the report. The length of the abstract can vary, but it should be under 150 words (approximately half a page of double spaced text). It serves very much like an **executive summary** in business reports. The abstract should contain information about the research question, data, findings, and conclusions of the study. The trick to creating a good abstract is using words as effectively as possible.

Do African Americans, Hispanics, and non-Hispanic whites differ in their explanations of the socioeconomic divide separating blacks and whites in the United States? Have such explanations changed over time? To answer these questions, I use data from the 1977 to 2004 General Social Surveys (GSS) to map race/ethnic differences in support for, trends in, and the determinants of seven "modes of explanation" for blacks' disadvantage. Trends over time indicate the continuation of a long-standing decline in non-Hispanic whites' use of an ability-based (innate inferiority) explanation. Non-Hispanic whites' beliefs in a purely motivational and a purely educational explanation are increasing, however, along with the view that none of the explanations offered in the GSS explain blacks' disadvantage. African Americans and Hispanics also evidence increases in a purely motivational explanation, but they differ from non-Hispanic whites in demonstrating clear declines in structural beliefs—especially the perception that discrimination explains blacks' lower socioeconomic status. These conservative shifts in blacks' and Hispanics' beliefs result in greater similarity with non-Hispanic whites over time. Notably, however, significant "static" race/ethnic group differences remain: non-Hispanic whites score highest, and blacks lowest, on a purely motivational explanation, while African Americans are more likely than both non-Hispanic whites and Hispanics to endorse a discrimination-based explanation. I conclude by discussing the implications of these findings for racial policy support.

Hunt, Matthew. 2007. "African American, Hispanic, and White Beliefs about Black/White Inequality, 1977-2004." *American Sociological Review*. 72: 390-415.

Figure 9.1 An Example of a Well Written Abstract

Figure 9.1 offers an example of a well-written abstract. A few aspects of this abstract deserve note. First, observe how the abstract starts with the research question and the data. It then states the findings and briefly explains the relationships (in order of magnitude) without using any statistics. The abstract concludes with a brief statement about the relevance of this research to our understandings of social relationships.

A similar approach should be used when creating an "executive summary." One could recast the above abstract into an executive summary using a "bulleted" format, as illustrated in Figure 9.2. Again, we suggest not writing the abstract or executive summary until all of the other sections of the research report have been written. Once the rest of the report is constructed, this section will fall into place quite readily.

African American, Hispanic, and White Beliefs about Black/White Inequality, 1977-2004
Executive Summary

Drawing on data from the General Social Surveys (GSS), this report examines the extent to which African Americans, Hispanics, and non-Hispanic whites differ in their explanations of the socioeconomic divide that separates blacks and whites in the United States. There are multiple explanations for the racial divides, some of which remain common, but others are in declining popularity. Key findings include:

A long-standing decline continues in non-Hispanic whites' use of an ability-based (innate inferiority) explanation, but the attributions of motivational and purely educational explanations to Black/White inequalities are increasing popular.

African Americans and Hispanics also are now more likely to express support for purely motivational explanations. But they also show declines in structural beliefs, which indicates that their explanations for racial differences are becoming more similar to those held by non-Hispanic whites.

But still, in 2004, African Americans differ from non-Hispanic whites and Hispanics, as they are more likely to endorse a discrimination-based explanation for Black/White Inequality.

Figure 9.2 An Abstract Translated into an Executive Summary

The Introduction

The **introduction** invites the reader to examine the rest of the report. It "sells" the report, informing the audience of the research question and why this question is important. It can be quite short (1-3 double spaced typed pages) and the length will depend on the type of research question addressed.

After reading through the introduction, the reader should have a very clear idea of the question the writer is addressing. Good introductions will often include the phrase "This paper examines the question…." This brief phrase forces the writer to come to terms with the specific purpose of the paper and to focus the rest of the report accordingly.

When writing an introduction, global and unsupported statements should be avoided. Consider this introductory sentence: "The growth in teenage births is a huge social problem." While this may seem intuitively appealing, the statement is factually incorrect because teenage births have declined rather than increased over the past 24 years. This introduction could be improved significantly by providing facts, such as "According to the Statistical Abstract of the United States, the teenage birth rate has declined from 69.5/1000 teenage women in 1970 to 41.2/1000 in 2004." The introduction can then build a compelling argument why it is still important to study teenage births in the context of current demographic trends.

Generally, the introduction should not serve as the conclusion of the study. Consider this introductory sentence: "In this paper I will prove that teenagers get pregnant because they lack solid moral values." It is important to remember that scientific methodology cannot prove anything; it can only find support for some hypotheses and refute other hypotheses. It would be a considerable improvement to rephrase this sentence as: "In this paper I will examine the degree to which personal values influence teenagers' risks for pregnancy." The introduction is designed only to open the research question, not to answer it.

The Literature Review

Because this book has been designed primarily to introduce data analysis strategies, we have not given much attention to the **literature review**. This is, however, a critical stage of the research process because it places the current study within a body of research. Science is a cumulative endeavor and all of scientists' work rests on the shoulders of preceding scientists. The literature review is aimed at simultaneously acknowledging other researchers' contributions to the collective knowledge, while also informing the reader of how these contributions relate to the current research project. The literature review is an appraisal of what is known and what is not known about the research question. Depending on the scope of the research project, typical literature reviews in journal articles range from 5-15 pages of double spaced type. For reports to professionals, the literature review will be briefer; for theses and dissertations, it is generally a chapter in length.

At a bare minimum, a literature review is an overview of the essential findings from articles related to the research question. However, creating a structure that links these articles will significantly improve a literature review. For example, in a study of factors associated with family violence, one of the authors organized the literature review by first writing about studies that looked at gender and family violence, then age and family violence, then race/ethnicity and family violence, then socioeconomic status and family violence. Providing a structure enables the reader to get a solid understanding of specific relationships (see Straus and Sweet 1992).

This is not the only way to structure a literature review. In some circumstances, it may make sense to link studies based on different methods. For example, a study on behaviors of the poor could first discuss participant-observation studies, then cross sectional survey studies, and then panel design studies. Each methodology will reveal different aspects of the behaviors of concern. In rare circumstances, it is appropriate to structure a literature review by the dates of the studies: for example, comparing studies on television viewing behavior performed in the 1950s with studies examining television viewing behavior in the 1990s.

By far, though, the weakest literature review discusses studies in random order or in the order in which the researcher read the articles. A writer should not forget the audience. The audience does not care about the sequence that the researcher read articles, or whether the researcher found the articles interesting or boring. They do care, however, about the research question, the knowledge that the scientific community has already developed, and how the current project fits into this body of literature.

The Methods

As with every other section, the **methods section** should be concise. It explains data collection, sampling strategies, sample sizes, indicators, and any reworking of the data. It is helpful to take the perspective of an intelligent and skeptical reader while writing the methods section. This reader looks for weaknesses and flaws in the study and will only trust conclusions if given reasons to. With this in mind, the intelligent and skeptical reader will ask questions such as:

- Are the data suited to the research question?
- Is the sample size large enough?
- Is the sample representative of the population?
- Is the sample appropriate to the study?
- Are the indicators reliable and valid?
- Does the researcher measure factors that could eliminate spurious findings?
- Does the researcher use analytic techniques suited to the data?

With each of these questions, the reader is asking, "To what degree will this research project suited to the research question?" The methods section builds that trust by carefully laying out the empirical basis of the study. It should, at a minimum, address the following information:

- The methods of data collection or sources of data
- The sample size and sample characteristics
- Modifications made to the data
- The types of statistical procedures used

The length of this section will vary, depending on the complexity of the data collection and analysis procedures, but generally, this section can be written in 2-5 double spaced typed pages. A report to a professional or administrative audience will often have a very brief methods section, but include relevant information in an appendix. Theses and dissertations tend to have an entire chapter dedicated to methodology.

The Findings

The **findings section** details the results of the study. Although the researchers may have found many interesting results, the findings should include only those results that directly relate to the research question posed in the introduction. This section includes graphs, tables, and text that describe relevant relationships.

There is an old maxim that social scientists read tables and ignore the text and "regular people" read the text and ignore the tables. There is truth to this statement and the findings section should be written accordingly. Tables and graphs should be rich with statistical information relating to the research question and should have enough information for the reader to interpret the table without having to simultaneously read the accompanying text. The text serves the purpose of translating these statistics into a form that "regular people" can understand. This involves describing the relationships without statistics. Tables may show percentages or regression coefficients, but the text accompanying the tables should be phrased in terms of "positive" and "negative" relationships.

One of the best ways to organize the findings section is to address results in relation to hypotheses. With some thought, you may be able to construct a modest number of tables that can effectively display all of the information needed to make conclusions, including significance tests, sample sizes, etc. Sometimes in the process of constructing tables, we observe patterns in relationships that show common tendencies among similar variable groupings. While these findings are created in the SPSS output, it is the composition of the findings into a chart form that oftentimes makes results most meaningful.

Below we offer some illustrations of different ways to compose tables, based on findings generated by different types of analytic procedures. Note that the tables illustrated below do not reproduce SPSS output. Rather, they take the key statistics from the output, and combine them into a digestible form.

Percentages of Faculty Dissatisfaction with Their Jobs: Full-Time Faculty Members by Sex

	Men		Women	
	All Full-Time Faculty	Assistant Professors	All Full-Time Faculty	Assistant Professors
Overall Dissatisfied With Job	15.2	17.9**	18.2	20.9**
Workload	30.4	34.7	35.5	39.8
Time to Work With Students	20.2	19.1	23.3	25.0
Time to Keep Current in Field	45.1	50.5	52.5	56.2
Time for Class Preparation	21.7	20.7	26/8	28.6

*p<.05 **p<.01 N=17,600. Adapted from Jacobs, Jerry and Sarah Winslow. 2004. "The Academic Life Course, Time Pressures and Gender Inequality." *Community, Work and Family*. 7: 143-161.

Figure 9.3 A Sample Table of That Summarizes a Series of Bivariate Analyses of Scale/Categorical Variables

Key to making tables like Figure 9.3 and Figure 9.4 is a structure that conveys a lot of information concisely. Notice how much information is included in each of these tables. The second table includes the logistic regression coefficients, standard errors, the constant, and significance tests for two different regressions. Academic journals commonly present the results of many regressions in a single table. But because many readers are not conversant in statistics, the data analyst is also commonly expected to translate these tables into prose.

Logistic Regression of Predictors of Verbal/Symbolic Aggression

	Verbal/Symbolic Aggression			
	Husband-to-Wife		Wife-to-Husband	
Variable	Coefficient	SE	Coefficient	SE
Gender (1=male)	0.20**	.077	0.39**	.077
Age	-0.02**	.003	-0.02**	.003
Number of children	-0.09**	.033	-0.05	.030
Days drunk	0.01**	.003	0.04**	.009
Days high on drugs	0.00	.001	0.01**	.002
Physical aggression	1.35**	.082	1.30**	.074
Couple conflict	0.93**	.052	0.83**	.050
Logit Constant	-1.82**	.162	-1.66**	.163

*p<.05 **p<.01 N=1662. Source: Straus, Murray and Stephen Sweet. 1992. "Verbal/Symbolic Aggression in Couples." Journal of Marriage and the Family 54:346-357.

Figure 9.4 A Sample Table Summarizing Two Separate Logistic Regressions

Depending on the complexities of the study, the findings section is typically 6-12 pages in a journal article. Findings will comprise the greatest amount of space in a report to a professional or administrative audience, and this audience will be better able to interpret graphs than tables. In theses and dissertations, findings can comprise many separate chapters.

The Conclusion

The **conclusion** relates the findings back to the research question posed in the introduction. The conclusion section should also discuss the limitations of the study and acknowledge any ways the data or analyses fall short of establishing causal relationships.

The conclusion reinforces the theoretical or practical implications of the findings. A good conclusion **generalizes** findings, relating the findings to social behavior outside of the study. Generalizing the findings requires consideration of **external validity**, the degree to which study findings accurately describe social behavior outside the confines of the study. In order to assess external validity, the researcher should consider the study's limitations. For example, laboratory experiments on human subjects often have little external validity because they place people in very contrived situations (Campbell and Stanley 1963). A biased sample also threatens external validity. If a sample consists of college freshman, for example, the researcher should be cautious in generalizing the findings to the general population.

One positive way to conclude a research report is to open the topic to additional related questions. The researcher can state "In this study I showed..." and then go on to say, "Further research needs to be done to find out if...." Science is a never-ending process and a strong conclusion acknowledges this.

The References

The **reference section** provides a list of sources cited in the report. Each discipline specifies different formats for referencing sources. It is important that the reference provide sufficient information so that a reader can locate the original source, no matter what the format.

The following are some examples of how the American Sociological Association suggests referencing books, articles, and government documents. Although other reference formats have different structures, they contain essentially the same information.

Books

Alba, Richard and Victor Nee. 2003. *Remaking the American Mainstream: Assimilation and Contemporary Immigration.* Cambridge, MA: Harvard University Press.

Articles from Collected Works

Altucher, Kristine and Lindy B. Williams. 2003. "Family Clocks: Timing Parenthood." Pp. 49-59 in *It's About Time: Career Strains, Strategies, and Successes*, edited by Phyllis Moen. Ithaca: Cornell University Press.

Articles in Journals

Waldinger, Roger, Nelson Lim, and David Cort. 2007. "Bad Jobs, Good Jobs, No Jobs? The Employment Experience of the Mexican-American Second Generation." *Journal of Ethnic and Migration Studies* 33:1-35

Articles from Newspapers and Magazines

Guiles, Melinda and Krystal Miller. 1990. "Mazda and Mitsubishi-Chrysler Venture Cut Output, Following Big Three's Lead." *Wall Street Journal*, January 12, pp. A2, A12.

Government Documents

U.S. Bureau of the Census. 2000. *Characteristics of Population.* Vol. 1. Washington, DC: U.S. Government Printing Office.

Summary

SPSS can generate a great variety of statistical output. The researcher's challenge is to organize and describe their statistical analysis so that a variety of audiences can understand it. The research report is essential in conveying this information. A well-written report is concise, conveys statistical information in numeric form, and describes relationships with prose. All good reports explain the relevance of the study, as well as acknowledge limitations of the study.

Key Terms

Abstract
Conclusion
Executive summary
Findings section
Introduction
Methods section
Literature review
Reference section
Research report
Title

Chapter 9 Exercises

One of the biggest challenges in writing is getting the pen to paper (or the fingers to the keyboard). In this chapter, we outlined different sections of the research report. The exercise for this chapter could involve selecting any one section of the report (ideally the Introduction, Literature Review, Methods, or Findings section) and begin writing.

If you want to write the Introduction or Literature Review, try to do so in a compelling way that highlights what may already be known about the question you are addressing. But beyond this, write about what we need to know, and how your project fits into this agenda. You could also perform the literature search, while trying to find articles that are recent, scholarly, and central to your research question.

If you are interested in writing about research methods, and are using the GSS04 data in your intended project (see chapter 10) you might try to compose this section and do some additional research on the General Social Survey to consider how those data were collected. You can also document in your report specific phrasings of key questions posed to participants in this study that will be included in your analysis. You can find much of this information at these websites:

http://webapp.icpsr.umich.edu/GSS/
http://www.norc.org/projects/General+Social+Survey.htm

If you want to start writing the Findings section, try to compose some tables that synthesize the output from the SPSS analyses, and also try describing the relationships revealed in these analyses. Composing tables can be a challenge, so pay attention to the approach other researchers have used in any articles you may review for your project.

We would not recommend attempting to write the Abstract or the Conclusion until all other parts of the report are drafted.

Chapter 10
Research Projects

Potential Research Projects

This concluding chapter describes some potential research projects that can be tackled with the data included with this book. Before embarking on these projects, it may be helpful to review some of the central lessons about the sequencing of data analysis efforts.

Research projects are guided by research questions and use data that can address these issues. Each of the projects we outline in this chapter are introduced by topical area. The first step in your project will be to shift these topics into sets of research questions. As you formulate these questions, you will need to evaluate the degree to which the STATES07 and the GSS04 data are suited to providing answers. This will involve locating variables that relate to these questions, finding relevant independent and dependent variables, and assessing the degree to which these variables accurately indicate of the concepts you want to examine. Inevitably, there will be some back and forth to finalize a set of questions that can be answered with the data that are available. While this may seem like a compromise, in fact many successful projects start out with the pragmatic concern of finding usable data. When the ideal data are not available, one is forced to "make-do" with existing resources, and sometime modify or even abandon questions initially proposed.

It will be to your advantage to understand what is already known about the research question that you are posing. Therefore, once you successfully meld your question with the data, it will be helpful to engage in a literature review on the subject. The depth of this literature review can be of your (or possibly your instructor's) discretion. At a minimum, you should read enough outside research to develop a reasonable understanding of what to anticipate in the theoretical framing of data analysis. Although this may seem like extra work, it will actually help push the analysis in fruitful directions more efficiently and effectively.

Empirical studies are hypothesis driven. Once you have developed an understanding of the relationships previously revealed by other researchers, you should be able to develop clearly framed hypotheses concerning the relationships between each independent and dependent variable. These hypotheses should be embedded in a larger theoretical understanding of the basis of social behavior and social experience.

Analysis of the data occurs in three stages. The first stage is univariate analysis, where you will examine the structures, central tendencies, and spreads of individual variables. The second stage is bivariate analysis, where you will examine the relationships between pairs of variables in isolation from other variables. The third stage is multivariate analysis, where multiple variables are examined together to control for extraneous factors and examine cumulative effects.

Once you analyze your data and understand the implications, you will be able to write the research report. The report distills the analysis to the most essential information so that readers can understand the relationships revealed by the data analysis. The report will contain carefully constructed tables and yield maximum content in minimal space. It also includes written descriptions of the observed relationships.

The report will conclude with a summary of supported and unsupported hypotheses. You will likely find that the data do not support all of your expectations. It is also important at this stage of the study to critically evaluate the limitations of the data. State level data, for instance, cover very large geographic regions. Therefore the data may gloss over important variations within each state. For example, we are writing this book in Tompkins County, a very rural part of New York State. Our experiences are likely very different than those of residents of Brooklyn County in New York City. However, because New York State, not the individual counties, is our unit of analysis in the STATES07 data, the variation between counties is lost. In some circumstances, lumping of social experience into such a large geographic area may limit the validity of the measures. This may be one consideration (among others) in evaluating the degree to which a theory is ultimately maintained or rejected. The research report should conclude with a discussion of these limitations, as well as with some suggestions for future research.

Finally, it is important to understand that analysis almost always involves some backtracking, reevaluation, restructuring of models, and recoding of variables. The best approach is an inquisitive one. Always think, is there a better or more robust way to test this relationship? Is there an alternate explanation for my findings? If there is, how can I test this? Data analysis, after all, is driven by a desire to understand the world that surrounds us. Enjoy!

Research Project 1: Racism

Social theory posits that people sometimes have prejudicial thoughts, which in turn lead them to discriminate against members of particular racial groups in society. This constitutes overt discrimination. Social theory also posits that there are structural factors that limit the opportunities of minority group members. Social structure includes the ways in which schools are funded, jobs distributed, infrastructure maintained, etc. If a racial group resides in an area that is deprived, they become deprived as a result. Can you determine, based on the GSS04 data and the STATES07 data, the degree to which prejudice, overt discrimination, and structural discrimination against African Americans persists in the United States?

Suggestions for Data Analysis:

A. Make a list of variables from the STATES07 data and the GSS04 data that relate to this research topic. Then form a research question that is suited to these data.

B. Determine which variables are independent variables and which variables are dependent variables.

C. Examine the structure of each of these variables to determine if they are numerical or categorical variables.

D. Develop a clear hypothesis as to how each independent variable will be related to the dependent variable, as informed by a literature review.

E. Perform Univariate Analysis
 Means, Medians, Modes, Standard Deviations, Box Plots, Histograms, Pie Charts, Bar Charts.

F. Perform Bivariate Analysis
 Crosstabulations, Comparison of Means, Correlations, Significance Tests, Box Plots, Bar Charts, Scatter Plots.

G. Perform Multivariate Analysis
 Linear Regression, Logistic Regression, Significance Tests, Scatter Plots.

Research Project 2: Suicide

What explains why a person chooses to terminate his or her own life? On the one hand, this can possibly be attributed to psychological factors, such as depression. On the other hand, it can be caused by social forces such as living in a society that is characterized as socially disorganized (e.g., lots of crime, unemployment, etc.). Can you then determine the degree to which social structural factors influence variation in suicide rates using the STATES07 data?

Suggestions for Data Analysis:

A. Make a list of variables from the STATES07 data that relate to this research topic. Then form a research question that is suited to these data.

B. Determine which variables are independent variables and which variables are dependent variables.

C. Examine the structure of each of these variables to determine if they are numerical or categorical variables.

D. Develop a clear hypothesis as to how each independent variable will be related to the dependent variable, as informed by a literature review.

E. Perform Univariate Analysis
Means, Medians, Modes, Standard Deviations, Box Plots, Histograms, Pie Charts, Bar Charts.

F. Perform Bivariate Analysis
Crosstabulations, Comparison of Means, Correlations, Significance Tests, Box Plots, Bar Charts, Scatter Plots.

G. Perform Multivariate Analysis
Linear Regression, Logistic Regression, Significance Tests, Scatter Plots

Research Project 3: Criminality

Some places have high crime rates and other places have low crime rates. What social factors contribute to this variation of crime from place to place? Do the methods of treating/punishing criminals have an effect on deterring crime? Use the STATES07 data to examine these issues. Pay particular attention to variation between types of crime in your analysis.

Suggestions for Data Analysis:

A. Make a list of variables from the STATES07 data that relate to this research topic. Then form a research question that is suited to these data.

B. Determine which variables are independent variables and which variables are dependent variables.

C. Examine the structure of each of these variables to determine if they are numerical or categorical variables.

D. Develop a clear hypothesis as to how each independent variable will be related to the dependent variable, as informed by a literature review.

E. Perform Univariate Analysis
Means, Medians, Modes, Standard Deviations, Box Plots, Histograms, Pie Charts, Bar Charts.

F. Perform Bivariate Analysis
Crosstabulations, Comparison of Means, Correlations, Significance Tests, Box Plots, Bar Charts, Scatter Plots.

G. Perform Multivariate Analysis
Linear Regression, Logistic Regression, Significance Tests, Scatter Plots

Research Project 4: Welfare and Other Public Aid Consumption

There is considerable variation in welfare use. One cause for this variation may be that individuals have different expectations of what they are entitled to receive just by living in our society. On the other hand, there may be social structural factors (e.g., unemployment) which either force or lure individuals to use welfare as a means of survival. Use the STATES07 data to examine which structural factors influence welfare consumption.

Suggestions for Data Analysis:

A. Make a list of variables from the STATES07 data which relate to this research topic. Then form a research question that is suited to these data.

B. Determine which variables are independent variables and which variables are dependent variables.

C. Examine the structure of each of these variables to determine if they are numerical or categorical variables.

D. Develop a clear hypothesis as to how each independent variable will be related to the dependent variable, as informed by a literature review.

E. Perform Univariate Analysis
Means, Medians, Modes, Standard Deviations, Box Plots, Histograms, Pie Charts, Bar Charts.

F. Perform Bivariate Analysis
Crosstabulations, Comparison of Means, Correlations, Significance Tests, Box Plots, Bar Charts, Scatter Plots.

G. Perform Multivariate Analysis
Linear Regression, Logistic Regression, Significance Tests, Scatter Plots

Research Project 5: Sexual Behavior

There is considerable variation in sexual behavior on an individual level, as well as by geographic locale. Using the GSS04 data, determine the social psychological factors that contribute to a permissive attitude toward sexual freedom. Or alternately, one can examine the potential ramifications of using pornography.

Suggestions for Data Analysis:

A. Make a list of variables from the GSS04 data that relate to this research topic. Then form a research question that is suited to these data.

B. Determine which variables are independent variables and which variables are dependent variables.

C. Examine the structure of each of these variables to determine if they are numerical or categorical variables.

D. Develop a clear hypothesis as to how each independent variable will be related to the dependent variable, as informed by a literature review.

E. Perform Univariate Analysis
Means, Medians, Modes, Standard Deviations, Box Plots, Histograms, Pie Charts, Bar Charts.

F. Perform Bivariate Analysis
Crosstabulations, Comparison of Means, Correlations, Significance Tests, Box Plots, Bar Charts, Scatter Plots.

G. Perform Multivariate Analysis
Linear Regression, Logistic Regression, Significance Tests, Scatter Plots

Research Project 6: Education

Individuals' educational attainment can be influenced by their personal background and the type of society in which they live. Using the GSS04 data, examine the degree to which educational attainment varies by a person's biographical background. Or, using the STATES07 data, examine the degree to which social structural factors influence educational attainment across the United States.

Suggestions for Data Analysis:

A. Make a list of variables from the STATES07 data and the GSS04 data that relate to this research topic. Then form a research question that is suited to these data.

B. Determine which variables are independent variables and which variables are dependent variables.

C. Examine the structure of each of these variables to determine if they are numerical or categorical variables.

D. Develop a clear hypothesis as to how each independent variable will be related to the dependent variable, as informed by a literature review.

E. Perform Univariate Analysis
Means, Medians, Modes, Standard Deviations, Box Plots, Histograms, Pie Charts, Bar Charts.

F. Perform Bivariate Analysis
Crosstabulations, Comparison of Means, Correlations, Significance Tests, Box Plots, Bar Charts, Scatter Plots.

G. Perform Multivariate Analysis
Linear Regression, Logistic Regression, Significance Tests, Scatter Plots

Research Project 7: Health

There are a variety of ways to measure health, including the incidence of specific ailments and their variation by place and demographic groupings. Mortality offers another indicator, as does the behavior of individuals and their health-related behaviors. Use the GSS04 and/or STATES07 data to assess the predictors of healthy lives.

Suggestions for Data Analysis:

A. Make a list of variables from the STATES07 data and the GSS04 data that relate to this research topic. Then form a research question that is suited to these data.

B. Determine which variables are independent variables and which variables are dependent variables.

C. Examine the structure of each of these variables to determine if they are numerical or categorical variables.

D. Develop a clear hypothesis as to how each independent variable will be related to the dependent variable, as informed by a literature review.

E. Perform Univariate Analysis
Means, Medians, Modes, Standard Deviations, Box Plots, Histograms, Pie Charts, Bar Charts.

F. Perform Bivariate Analysis
Crosstabulations, Comparison of Means, Correlations, Significance Tests, Box Plots, Bar Charts, Scatter Plots.

G. Perform Multivariate Analysis
Linear Regression, Logistic Regression, Significance Tests, Scatter Plots

Research Project 8: Happiness

What predicts happiness – Money? Life stage? Marital status? Gender? These are only a few of the questions you can examine with the GSS04 data. Can you develop a theory about what causes happiness and then test the associations revealed in the data?

Suggestions for Data Analysis:

A. Make a list of variables from the GSS04 data that relate to this research topic. Then form a research question that is suited to these data.

B. Determine which variables are independent variables and which variables are dependent variables.

C. Examine the structure of each of these variables to determine if they are numerical or categorical variables.

D. Develop a clear hypothesis as to how each independent variable will be related to the dependent variable, as informed by a literature review.

E. Perform Univariate Analysis
Means, Medians, Modes, Standard Deviations, Box Plots, Histograms, Pie Charts, Bar Charts.

F. Perform Bivariate Analysis
Crosstabulations, Comparison of Means, Correlations, Significance Tests, Box Plots, Bar Charts, Scatter Plots.

G. Perform Multivariate Analysis
Linear Regression, Logistic Regression, Significance Tests, Scatter Plots

Research Project 9: Your Topic

There are many other variables included in the STATES07 data set and the GSS04 data set. Can you develop a research question that uses these data to their fullest potential? Ultimately you may need to search for additional data to merge with the STATES07 data, but don't let this stop you. That is to be expected in social research.

Suggestions for Data Analysis:

A. Make a list of variables from the STATES07 data and the GSS04 data that relate to this research topic. Then form a research question that is suited to these data.

B. Determine which variables are independent variables and which variables are dependent variables.

C. Examine the structure of each of these variables to determine if they are numerical or categorical variables.

D. Develop a clear hypothesis as to how each independent variable will be related to the dependent variable, as informed by a literature review.

E. Perform Univariate Analysis
Means, Medians, Modes, Standard Deviations, Box Plots, Histograms, Pie Charts, Bar Charts.

F. Perform Bivariate Analysis
Crosstabulations, Comparison of Means, Correlations, Significance Tests, Box Plots, Bar Charts, Scatter Plots.

G. Perform Multivariate Analysis
Linear Regression, Logistic Regression, Significance Tests, Scatter Plots

APPENDIX 1: STATES07 DESCRIPTIVES

Descriptive Statistics

	N	Mean
DMS378 Birth Rate per 1,000 Pop.: 2005	51	13.778
DMS379 Births of Low Birthweight as a Percent of All Births: 2005	51	8.149
DMS380 Teenage Birth Rate per 1,000 Women Age 15-19: 2005	51	40.406
DMS381 Births to Unmarried Women as a Percent of All Births: 2005	51	36.394
DMS408 Persons per Household: 2005	51	2.538
DMS410 Percent of Households Headed by Married Couples: 2005	51	49.971
DMS411 Percent of Households Headed by Single Mothers: 2005	51	7.333
DMS429 Population: 2006	51	5870559
DMS434 Population per Square Mile: 2006	51	373.265
DMS439 Percent of Population White: 2005	51	81.537
DMS441 Percent of Population Black: 2005	51	11.278
DMS443 Percent of Population Hispanic: 2005	51	9.027
DMS445 Percent of Population Asian: 2005	51	3.343
DMS447 Percent of Population American Indian: 2005	51	1.743
DMS449 Percent of Population of Mixed Race: 2005	51	1.808
DMS454 Median Age: 2005	51	36.998
DMS468 Percent of Population 65 Years Old and Older: 2005	51	12.604
DMS474 Percent of Population Foreign Born: 2005	51	8.184
DMS475 Percent of Population Speaking a Language Other Than English at Home: 2005	51	13.337
DMS476 Percent of Population Speaking Spanish at Home: 2005	51	7.310
DMS478 Marriage Rate per 1,000 Pop.: 2005	51	8.659
DMS483 Divorce Rate per 1,000 Pop.: 2005	46	3.802
DMS484 Average Family Size: 2005	51	3.087
PVS493 Poverty Rate: 2005	51	12.076
PVS495 Percent of Children Living in Poverty: 2005	51	17.590
PVS497 Percent of Female-Headed Families with Children Living in Poverty: 2005	51	38.020
PVS518 Percent of Population Receiving Public Aid: 2004	51	3.706
PVS522 Average Monthly TANF Assistance per Recipient: 2004	51	139.768
PVS527 Percent of Population Receiving Food Stamps: 2006	51	9.153
CRC59 White State Prisoner Incarceration Rate per 100,000: 2005	49	407.898
CRC60 Black State Prisoner Incarceration Rate per 100,000: 2005	49	2541.857
CRC61 Hispanic State Prisoner Incarceration Rate per 100,000: 2005	40	702.900
CRC180 Reported Juvenile Arrest Rate per 100,000: 2005	49	6835.788
CRC186 Reported Juvenile Arrest Rate per 100,000 for Violent Crime: 2005	49	234.012
CRC201 Reported Juvenile Arrest Rate per 100,000 for Property Crime: 2005	49	1371.035
CRC222 Reported Juvenile Arrest Rate per 100,000 for Drug Abuse Violations: 2005	49	537.518
CRC248 Rate per 1,000 of Child Abuse and Neglect: 2004	50	11.500
CRS26 Crimes: 2005	51	226605.0
CRS28 Crime Rate per 100,000: 2005	51	3799.533
CRS30 Violent Crimes: 2005	51	27268.529
CRS32 Violent Crime Rate per 100,000: 2005	51	422.343
CRS34 Murders: 2005	51	327.294
CRS35 Murder Rate per 100,000: 2005	51	5.316

	N	Mean
CRS37 Rapes: 2005	51	1841.843
CRS38 Rape Rate per 100,000: 2005	51	33.729
CRS39 Robberies: 2005	51	8178.863
CRS40 Robbery Rate per 100,000: 2005	51	113.918
CRS41 Aggravated Assaults: 2005	51	16920.529
CRS42 Aggravated Assault Rate per 100,000: 2005	51	269.388
CRS43 Property Crimes: 2005	51	199336.5
CRS45 Property Crime Rate per 100,000: 2005	51	3377.190
CRS47 Burglaries: 2005	51	42237.765
CRS48 Burglary Rate per 100,000: 2005	51	690.500
CRS49 Larcenies and Thefts: 2005	51	132878.6
CRS50 Larceny and Theft Rate per 100,000: 2005	51	2303.486
CRS51 Motor Vehicle Thefts: 2005	51	24220.118
CRS52 Motor Vehicle Theft Rate per 100,000: 2005	51	383.216
CRS55 Reported Arrest Rate per 100,000: 2005	49	5016.137
CRS57 Prisoners in State Correctional Institutions: Year End 2005	50	26766.120
CRS58 State Prisoner Incarceration Rate per 100,000: 2005	50	400.640
CRS60 Prisoners Under Sentence of Death: 2005	38	84.658
CRC77 Prisoners Executed: 1977 to 2005	51	19.627
CRS61 Annual Operating Costs per State Prisoner: 2001	51	24104.039
DFS65 Homeland Security Grants: 2006	51	3E+007
DFS66 Per Capita Homeland Security Grants: 2006	51	8.063
DFS74 U.S. Department of Defense Active Duty Military Personnel: 2005	51	22417.686
DFS75 U.S. Department of Defense Domestic Civilian Personnel: 2005	51	12534.373
DFS76 U.S. Department of Defense Reserve and National Guard Personnel: 2005	51	20886.804
DFS83 Percent of Adult Population Who are Veterans: 2006	51	11.380
DFS85 Rate per 100,000 of U.S. Military Fatalities in Iraq as of January 25, 2007	51	1.180
ECS424 Homeownership Rate: 2005	51	70.278
ECS89 Per Capita Gross Domestic Product: 2005	51	42740.431
ECS93 Per Capita Personal Income: 2005	51	33739.176
ECS96 Median Household Income: 2005	50	46098.360
ECS98 Personal Bankruptcy Rate per 100,000: 2006	51	361.569
EDS115 Percent of Elem./Second. School Students in Private Schools: 2004	51	9.522
EDS117 Average Class Size in Public Elem. Schools: 2000	51	20.649
EDS118 Average Class Size in Public Second. Schools: 2000	51	22.680
EDS120 Estimated Pupil-Teacher Ratio in Public Elem. and Second. Schools: 2006	51	15.216
EDS121 Estimated Average Salary of Public School Classroom Teachers: 2006	51	46593.922
EDS123 Percent of Public School 4th Graders Proficient or Better in Reading: 2005	51	30.490
EDS124 Percent of Public School 8th Graders Proficient or Better in Reading: 2005	51	29.745
EDS125 Percent of Public School 4th Graders Proficient or Better in Math: 2005	51	35.176
EDS126 Percent of Public School 8th Graders Proficient or Better in Math: 2005	50	28.400
EDS128 Estimated Public High School Graduation Rate: 2006	51	72.014
EDS129 Percent of Population Graduated from High School: 2005	51	85.649
EDS130 Public High School Drop Out Rate: 2002	46	4.404

	N	Mean
EDS131 ACT Average Composite Score: 2006	51	21.302
EDS133 Per Capita State and Local Govt. Spending for Education: 2004	51	2231.569
EDS136 Per Capita State and Local Govt. Spending for Elem. and Second. Education: 2004	51	1503.000
EDS140 Per Capita State and Local Govt. Spending for Higher Education: 2004	51	622.882
EDS147 Enrollment Rate per 1,000 Aged 18-24 in Institutions of Higher Education: 2004	51	601.647
EDS150 Percent of Population With a Bachelors Degree or More: 2005	51	27.357
EMS161 Median Earnings of Male Full-Time Workers: 2005	51	41702.961
EMS162 Median Earnings of Female Full-Time Workers: 2005	51	31564.824
EMS163 State Minimum Wage Rates: 2007	46	6.195
EMS177 Percent of Women in the Civilian Labor Force: 2004	51	60.682
EMS179 Percent of Children Under 6 Years Old With All Parents Working: 2005	51	62.061
ENS219 Hazardous Waste Sites on the National Priority List: 2006	51	25.196
ENS222 Toxic Releases: Total Pollution Released: 2004	51	8E+007
ENS223 Toxic Releases: Total Air Emissions: 2004	51	3E+007
ENS224 Toxic Releases: Total Surface Water Discharges: 2004	51	4719164
HTC244 Percent of High School Students Who Drink Alcohol: 2005	40	42.355
HTC245 Percent of High School Students Who Use Marijuana: 2005	40	19.445
HTH98 Infant Mortality Rate per 1,000 Live Births: 2003	51	6.951
HTH100 White Infant Mortality Rate per 1,000 Live Births: 2003	50	5.844
HTH102 Black Infant Mortality Rate per 1,000 Live Births: 2003	36	14.203
HTH167 Death Rate per 100,000 by Accidents: 2003	51	41.345
HTH170 Death Rate per 100,000 by Motor Vehicle Accidents: 2003	51	17.073
HTH173 Death Rate per 100,000 by Firearm Injury: 2003	51	11.298
HTH176 Death Rate per 100,000 by Homicide: 2003	46	6.217
HTH179 Death Rate per 100,000 by Suicide: 2003	51	12.227
HTH182 Death Rate per 100,000 by Alcohol-Induced Deaths: 2003	51	7.961
HTH185 Occupational Fatality Rate per 100,000: 2005	51	4.782
HTH386 Sexually Transmitted Disease Rate per 100,000: 2005	51	434.808
HTH393 Percent of Adults Who Have Asthma: 2005	51	8.094
HTH394 Percent of Children Who Have Asthma: 2005	42	8.679
HTH397 Percent of Adults Reporting Serious Psychological Distress: 2004	51	9.822
HTH488 Adult Per Capita Beer Consumption: 2004	51	32.012
HTH490 Adult Per Capita Wine Consumption: 2004	51	3.053
HTH492 Adult Per Capita Distilled Spirits Consumption: 2004	51	2.059
HTH493 Percent of Adults Who Do Not Drink Alcohol: 2005	51	46.790
HTH495 Percent of Adults Who Smoke: 2005	51	20.988
HTH500 Percent of Population Who are Illicit Drug Users: 2004	51	8.208
HTH504 Percent of Adults Who Do Not Exercise: 2005	51	24.382
HTH505 Percent of Adults Who Exercise Vigorously: 2005	51	27.486
HTH507 Percent of Adults with High Blood Pressure: 2005	51	25.786
HTH509 Percent of Adults Who Have Visited a Dentist or Dental Clinic: 2004	50	70.734
HTH510 Percent of Adults 65 and Older Who Have Lost All Their Natural Teeth: 2004	50	22.186

	N	Mean
HTH512 Percent of Adults Rating Their Health as Fair or Poor: 2005	51	15.637
HTS335 Tornadoes: 2005	51	91.724
HTS367 Percent of Private-Sector Establishments That Offer Health Insurance: 2004	51	54.125
HTS369 Percent of Population Not Covered by Health Insurance: 2005	51	14.445
HTS370 Percent of Population Lacking Access to Primary Care: 2006	51	12.657
HTS372 Rate per 100,000 Pop. of Physicians: 2005	51	299.569
HTS373 Rate per 100,000 Pop. of Registered Nurses: 2005	51	845.608
HTS374 Rate per 100,000 Pop. of Dentists: 2004	51	57.863
HTS382 Percent of Mothers Receiving Late or No Prenatal Care: 2004	42	3.564
HTS384 Reported Legal Abortions per 1,000 Live Births: 2003	47	220.638
HTS386 Infant Mortality Rate per 1,000 Live Births: 2005	51	6.882
HTS388 Age-Adjusted Death Rate per 100,000 Pop.: 2004	48	814.746
HTS390 Estimated Death Rate per 100,000 Pop. by Cancer: 2007	51	192.625
HTS392 Estimated Rate per 100,000 Pop. of New Cancer Cases: 2007	51	489.702
HTS394 Age-Adjusted Death Rate per 100,000 Pop. by Accidents: 2003	51	40.933
HTS398 Age-Adjusted Death Rate per 100,000 Pop. by Diseases of the Heart: 2003	51	226.465
HTS400 Age-Adjusted Death Rate per 100,000 Pop. by Suicide: 2003	51	12.200
HTS401 Death Rate per 100,000 Pop. by AIDS: 2003	40	5.178
HTS402 Age-Adjusted Death Rate per 100,000 Pop. by AIDS: 2003	40	5.195
HTS403 Adult Per Capita Alcohol Consumption: 2004	51	2.675
HTS404 Percent of Adults Who Smoke: 2005	51	20.988
HTS405 Percent of Adults Overweight or Obese: 2005	51	61.029
HTS406 Percent of Children Aged 19 to 35 Months Fully Immunized: 2005	51	75.006
HTS547 Highway Fatality Rate per 100 Million Vehicle Miles Traveled: 2005	51	1.510
HTS550 Safety Belt Usage Rate: 2006	42	81.581
LES62 Per Capita State & Local Govt. Spending for Police Protection: 2004	51	221.941
LES63 Per Capita State and Local Govt. Spending for Corrections: 2004	51	175.941
LES64 Rate per 100,000 of Full-Time Sworn Officers in Law Enforcement Agencies: 2000	51	245.765
POS490 Percent of Eligible Voters Reported Registered: 2004	51	73.339
POS492 Percent of Eligible Population Reported Voting: 2004	51	65.047

APPENDIX 2: GSS04.SAV FILE INFORMATION

```
Name (Position) Label

YEAR (1) GSS YEAR FOR THIS RESPONDENT
           Measurement level: Scale

ID (2) RESPONDNT ID NUMBER
           Measurement level: Scale

MARITAL (3) MARITAL STATUS
           Measurement level: Nominal
           Missing Values: 9
           Value    Label
               1    MARRIED
               2    WIDOWED
               3    DIVORCED
               4    SEPARATED
               5    NEVER MARRIED
               9 M  NA

AGE (4) AGE OF RESPONDENT
           Measurement level: Scale
           Missing Values: 98, 99
           Value    Label
              98 M  DK
              99 M  NA

EDUC (5) HIGHEST YEAR OF SCHOOL COMPLETED
           Measurement level: Scale
           Missing Values: 98, 99
           Value    Label
              98 M  DK
              99 M  NA

DEGREE (6) RS HIGHEST DEGREE
           Measurement level: Ordinal
           Missing Values: 8, 9
           Value    Label
               0    LT HIGH SCHOOL
               1    HIGH SCHOOL
               2    JUNIOR COLLEGE
               3    BACHELOR
               4    GRADUATE
               8 M  DK
               9 M  NA

SEX (7) RESPONDENTS SEX
           Measurement level: Nominal
           Value    Label
               0    FEMALE
               1    MALE

RACE (8) RACE OF RESPONDENT
           Measurement level: Nominal
           Missing Values: 0
```

```
          Value     Label
              0 M   NAP
              1     WHITE
              2     BLACK
              3     OTHER

RACECEN1 (9) WHAT IS RS RACE 1ST MENTION
          Measurement level: Nominal
          Missing Values: 98, 99
          Value     Label
              1     WHITE
              2     BLACK OR AFRICAN AMERICAN
              3     AMERICAN INDIAN OR ALASKA NATIVE
              4     ASIAN INDIAN
              5     CHINESE
              6     FILIPINO
              7     JAPANESE
              8     KOREAN
              9     VIETNAMESE
             10     OTHER ASIAN
             11     NATIVE HAWAIIAN
             12     GUAMANIAN OR CHAMORRO
             13     SAMOAN
             14     OTHER PACIFIC ISLANDER
             15     SOME OTHER RACE
             16     HISPANIC
             98 M   DK
             99 M   NA

RACECEN2 (10) WHAT IS RS RACE 2ND MENTION
          Measurement level: Nominal
          Missing Values: 98, 99
          Value     Label
              1     WHITE
              2     BLACK OR AFRICAN AMERICAN
              3     AMERICAN INDIAN OR ALASKA NATIVE
              4     ASIAN INDIAN
              5     CHINESE
              6     FILIPINO
              7     JAPANESE
              8     KOREAN
              9     VIETNAMESE
             10     OTHER ASIAN
             11     NATIVE HAWAIIAN
             12     GUAMANIAN OR CHAMORRO
             13     SAMOAN
             14     OTHER PACIFIC ISLANDER
             15     SOME OTHER RACE
             16     HISPANIC
             98 M   DK
             99 M   NA
```

```
RACECEN3 (11) WHAT IS RS RACE 3RD MENTION
          Measurement level: Nominal
          Missing Values: 98, 99
          Value    Label
              1    WHITE
              2    BLACK OR AFRICAN AMERICAN
              3    AMERICAN INDIAN OR ALASKA NATIVE
              4    ASIAN INDIAN
              5    CHINESE
              6    FILIPINO
              7    JAPANESE
              8    KOREAN
              9    VIETNAMESE
             10    OTHER ASIAN
             11    NATIVE HAWAIIAN
             12    GUAMANIAN OR CHAMORRO
             13    SAMOAN
             14    OTHER PACIFIC ISLANDER
             15    SOME OTHER RACE
             16    HISPANIC
             98 M  DK
             99 M  NA

COHORT (12) YEAR OF BIRTH
          Measurement level: Scale
          Missing Values: 0, 9999
          Value    Label
              0 M  NAP
           9999 M  NA

ZODIAC (13) RESPONDENTS ASTROLOGICAL SIGN
          Measurement level: Nominal
          Missing Values: 98, 99
          Value    Label
              1    ARIES
              2    TAURUS
              3    GEMINI
              4    CANCER
              5    LEO
              6    VIRGO
              7    LIBRA
              8    SCORPIO
              9    SAGITTARIUS
             10    CAPRICORN
             11    AQUARIUS
             12    PISCES
             98 M  DK
             99 M  NA

DIVORCE (14) EVER BEEN DIVORCED OR SEPARATED
          Measurement level: Nominal
          Missing Values: 9
          Value    Label
              0    NO
              1    YES
```

```
WIDOWED (15) EVER BEEN WIDOWED
           Measurement level: Nominal
           Missing Values: 9
           Value    Label
               0    NO
               1    YES

SIBS (16) NUMBER OF BROTHERS AND SISTERS
           Measurement level: Scale
           Missing Values: 98, 99
           Value    Label
              98 M  DK
              99 M  NA

CHILDS (17) NUMBER OF CHILDREN
           Measurement level: Scale
           Missing Values: 9
           Value    Label
               0    NONE
               1    ONE
               2    TWO
               3    THREE
               4    FOUR
               5    FIVE
               6    SIX
               7    SEVEN
               8    EIGHT OR MORE
               9 M  NA

AGEKDBRN (18) R'S AGE WHEN 1ST CHILD BORN
           Measurement level: Scale
           Missing Values: 98, 99
           Value    Label
              98 M  DK
              99 M  NA

HOMPOP (19) NUMBER OF PERSONS IN HOUSEHOLD
           Measurement level: Scale
           Missing Values: 98, 99
           Value    Label
              98 M  DK
              99 M  NA

BABIES (20) HOUSEHOLD MEMBERS LESS THAN 6 YRS OLD
           Measurement level: Scale
           Missing Values: 9
           Value    Label
               8    8 OR MORE
               9 M  NA

PRETEEN (21) HOUSEHOLD MEMBERS 6 THRU 12 YRS OLD
           Measurement level: Scale
           Missing Values: 9
           Value    Label
               8    8 OR MORE
               9 M  NA
```

```
TEENS (22  HOUSEHOLD MEMBERS 13 THRU 17 YRS OLD
           Measurement level: Scale
           Missing Values: 9
           Value    Label
               8    8 OR MORE
               9 M  NA

ADULTS (23) HOUSEHOLD MEMBERS 18 YRS AND OLDER
           Measurement level: Scale
           Missing Values: 9
           Value    Label
               8    8 OR MORE
               9 M  NA

UNRELAT (24) NUMBER IN HOUSEHOLD NOT RELATED
           Measurement level: Scale
           Missing Values: -1, 9
           Value    Label
              -1 M  NAP
               8    8 OR MORE
               9 M  NA

RESPNUM (25) NUMBER IN FAMILY OF R
           Measurement level: Scale
           Missing Values: 0, 99
           Value    Label
               0 M  NAP
               1    1ST PERSON
               2    2ND PERSON
               3    3RD PERSON
               4    4TH PERSON
               5    5TH PERSON
               6    6TH PERSON
               7    7TH PERSON
               8    8TH PERSON
               9    9TH PERSON
              10    10TH PERSON
              11    11TH PERSON
              12    12TH PERSON
              13    13TH PERSON
              14    14TH PERSON
              99 M  NA
```

```
HHTYPE1 (26) HOUSEHOLD TYPE (CONDENSED)
           Measurement level: Nominal
           Missing Values: 0, 99
           Value    Label
               0 M  NAP
               1    MARRIED COUPLE, NO CHILDREN
               2    SINGLE PARENT
               3    OTHER FAM., NO CHILDREN
               4    SINGLE ADULT
               5    COHAB COUPLE, NO CHILDREN
               6    NON-FAMILY, NO CHILDREN
               8    UNSURE, NO CHILDREN
              11    MARRIED COUPLE W CHILDREN
              13    OTHER FAMILY W CHILDREN
              15    COHAB COUPLE W CHILDREN
              16    NON-FAMILY W CHILDREN
              18    UNSURE W CHILDREN
              99 M  NA

PRESTG80 (27) RS OCCUPATIONAL PRESTIGE SCORE (1980)
           Measurement level: Scale
           Missing Values: 0
           Value    Label
               0 M  DK,NA,NAP

SPPRES80 (28) SPOUSES OCCUPATIONAL PRESTIGE SCORE (1980)
           Measurement level: Scale
           Missing Values: 0
           Value    Label
               0 M  DK,NA,NAP

INCOME98 (29) TOTAL FAMILY INCOME
           Measurement level: Ordinal
           Missing Values: 24, 98, 99
           Value    Label
               1    UNDER $1 000
               2    $1 000 TO 2 999
               3    $3 000 TO 3 999
               4    $4 000 TO 4 999
               5    $5 000 TO 5 999
               6    $6 000 TO 6 999
               7    $7 000 TO 7 999
               8    $8 000 TO 9 999
               9    $10000 TO 12499
              10    $12500 TO 14999
              11    $15000 TO 17499
              12    $17500 TO 19999
              13    $20000 TO 22499
              14    $22500 TO 24999
              15    $25000 TO 29999
              16    $30000 TO 34999
              17    $35000 TO 39999
              18    $40000 TO 49999
              19    $50000 TO 59999
              20    $60000 TO 74999
              21    $75000 - $89999
              22    $90000- $109999
```

```
              23    $110 000 OVER
              24 M  REFUSED
              98 M  DK
              99 M  NA

RINCOME98 (30) RESPONDENTS INCOME
          Measurement level: Ordinal
          Missing Values: 24, 98, 99
          Value    Label
               1    UNDER $1 000
               2    $1 000 TO 2 999
               3    $3 000 TO 3 999
               4    $4 000 TO 4 999
               5    $5 000 TO 5 999
               6    $6 000 TO 6 999
               7    $7 000 TO 7 999
               8    $8 000 TO 9 999
               9    $10000 TO 12499
              10    $12500 TO 14999
              11    $15000 TO 17499
              12    $17500 TO 19999
              13    $20000 TO 22499
              14    $22500 TO 24999
              15    $25000 TO 29999
              16    $30000 TO 34999
              17    $35000 TO 39999
              18    $40000 TO 49999
              19    $50000 TO 59999
              20    $60000 TO 74999
              21    $75000 - $89999
              22    $90000- $109999
              23    $110 000 OVER
              24 M  REFUSED
              98 M  DK
              99 M  NA

EARNRS (31) HOW MANY IN FAMILY EARNED MONEY
          Measurement level: Scale
          Missing Values: 9
          Value    Label
               8    EIGHT OR MORE
               9 M  NA

WRKSTAT (32) LABOR FRCE STATUS
          Measurement level: Nominal
          Missing Values: 0, 9
          Value    Label
               0 M  NAP
               1    WORKING FULLTIME
               2    WORKING PARTTIME
               3    TEMP NOT WORKING
               4    UNEMPL, LAID OFF
               5    RETIRED
               6    SCHOOL
               7    KEEPING HOUSE
               8    OTHER
               9 M  NA
```

```
HRS1 (33) NUMBER OF HOURS WORKED LAST WEEK
          Measurement level: Scale
          Missing Values: 98, 99
          Value    Label
             98 M  DK
             99 M  NA

WRKSLF (34) R SELF-EMP OR WORKS FOR SOMEBODY
          Measurement level: Nominal
          Missing Values: 9
          Value    Label
              0    SOMEONE ELSE
              1    SELF-EMPLOYED

PARTFULL (35) WAS R'S WORK PART-TIME OR FULL-TIME?
          Measurement level: Nominal
          Value    Label
              0    PART-TIME
              1    FULL-TIME

SPWRKSTA (36) SPOUSE LABOR FORCE STATUS
          Measurement level: Nominal
          Missing Values: 0, 9
          Value    Label
              0 M  NAP
              1    WORKING FULLTIME
              2    WORKING PARTTIME
              3    TEMP NOT WORKING
              4    UNEMPL, LAID OFF
              5    RETIRED
              6    SCHOOL
              7    KEEPING HOUSE
              8    OTHER
              9 M  NA

SPHRS1 (37) NUMBER OF HRS SPOUSE WORKED LAST WEEK
          Measurement level: Scale
          Missing Values: -1, 99
          Value    Label
             98    DK
             99 M  NA

SPEVWORK (38) SPOUSE EVER WORK AS LONG AS A YEAR
          Measurement level: Nominal
          Missing Values: 9
          Value    Label
              0    NO
              1    YES
```

```
DWELOWN (39) DOES R OWN OR RENT HOME?
          Measurement level: Nominal
          Missing Values: 8, 9
          Value    Label
              1    OWN OR IS BUYING
              2    PAYS RENT
              3    OTHER
              8 M  DK
              9 M  NA

SEI (40) RESPONDENT SOCIOECONOMIC INDEX
          Measurement level: Scale
          Missing Values: .0, 99.8, 99.9
          Value    Label
             .0 M  NAP
           99.8 M  DK
           99.9 M  NA

SPSEI (41) R'S SPOUSE'S SOCIOECONOMIC INDEX
          Measurement level: Scale
          Missing Values: .0, 99.8, 99.9
          Value    Label
             .0 M  NAP
           99.8 M  DK
           99.9 M  NA

HAPMAR (42) HAPPINESS OF MARRIAGE
          Measurement level: Ordinal
          Missing Values: 8, 9
          Value    Label
              1    VERY HAPPY
              2    PRETTY HAPPY
              3    NOT TOO HAPPY
              8 M  DK
              9 M  NA

DIVLAW (43) DIVORCE LAWS
          Measurement level: Ordinal
          Missing Values: 8, 9
          Value    Label
              1    EASIER
              2    MORE DIFFICULT
              3    STAY SAME
              8 M  DK
              9 M  NA

SPANKING (44) FAVOR SPANKING TO DISCIPLINE CHILD
          Measurement level: Ordinal
          Missing Values: 8, 9
          Value    Label
              1    STRONGLY AGREE
              2    AGREE
              3    DISAGREE
              4    STRONGLY DISAGREE
              8 M  DK
              9 M  NA
```

```
MARHOMO (45) HOMOSEXUALS SHOULD HAVE RIGHT TO MARRY
           Measurement level: Ordinal
           Missing Values: 8, 9
           Value    Label
               1    STRONGLY AGREE
               2    AGREE
               3    NEITHER AGREE NOR DISAGREE
               4    DISAGREE
               5    STRONGLY DISAGREE
               8 M  CAN T CHOOSE
               9 M  NA

HAPPY (46) GENERAL HAPPINESS
           Measurement level: Ordinal
           Missing Values: 8, 9
           Value    Label
               1    VERY HAPPY
               2    PRETTY HAPPY
               3    NOT TOO HAPPY
               8 M  DK
               9 M  NA

HEALTH (47) CONDITION OF HEALTH
           Measurement level: Ordinal
           Missing Values: 8, 9
           Value    Label
               1    EXCELLENT
               2    GOOD
               3    FAIR
               4    POOR
               8 M  DK
               9 M  NA

LIFE (48) IS LIFE EXCITING OR DULL
           Measurement level: Ordinal
           Missing Values: 8, 9
           Value    Label
               1    EXCITING
               2    ROUTINE
               3    DULL
               8 M  DK
               9 M  NA

LETDIE1 (49) ALLOW INCURABLE PATIENTS TO DIE
           Measurement level: Nominal
           Missing Values: 9
           Value    Label
               0    NO
               1    YES
```

```
RUSHED (50) HOW OFTEN R FEELS RUSHED
          Measurement level: Ordinal
          Missing Values: 8, 9
          Value    Label
              1    ALWAYS
              2    SOMETIMES
              3    ALMOST NEVER
              8 M  DK
              9 M  NA

HEALTH1 (51) RS HEALTH IN GENERAL
          Measurement level: Ordinal
          Missing Values: 8, 9
          Value    Label
              1    Excellent
              2    Very good
              3    Good
              4    Fair
              5    Poor
              8 M  DONT KNOW
              9 M  NO ANSWER

MNTLHLTH (52) DAYS OF POOR MENTAL HEALTH PAST 30 DAYS
          Measurement level: Scale
          Missing Values: -1, 98, 99
          Value    Label
             -1 M  NAP
             98 M  DONT KNOW
             99 M  NO ANSWER

SATSELF (53) ON THE WHOLE I AM SATISFIED WITH MYSELF
          Measurement level: Ordinal
          Missing Values: 0, 8, 9
          Value    Label
              0 M  NAP
              1    STRONGLY AGREE
              2    AGREE
              3    DISAGREE
              4    STRONGLY DISAGREE
              8 M  DONT KNOW
              9 M  NO ANSWER

AFAILURE (54) I AM INCLINED TO FEEL I AM A FAILURE
          Measurement level: Ordinal
          Missing Values: 0, 8, 9
          Value    Label
              0 M  NAP
              1    STRONGLY AGREE
              2    AGREE
              3    DISAGREE
              4    STRONGLY DISAGREE
              8 M  DONT KNOW
              9 M  NO ANSWER
```

```
SLFRSPCT (55) I WISH I COULD HAVE MORE RESPECT FOR MYSELF
           Measurement level: Ordinal
           Missing Values: 0, 8, 9
           Value    Label
               0 M  NAP
               1    STRONGLY AGREE
               2    AGREE
               3    DISAGREE
               4    STRONGLY DISAGREE
               8 M  DONT KNOW
               9 M  NO ANSWER

OFWORTH (56) I AM A PERSON OF WORTH AT LEAST EQUAL TO OTHERS
           Measurement level: Ordinal
           Missing Values: 0, 8, 9
           Value    Label
               0 M  NAP
               1    STRONGLY AGREE
               2    AGREE
               3    DISAGREE
               4    STRONGLY DISAGREE
               8 M  DONT KNOW
               9 M  NO ANSWER

NOGOOD (57) AT TIMES I THINK I AM NO GOOD AT ALL
           Measurement level: Ordinal
           Missing Values: 0, 8, 9
           Value    Label
               0 M  NAP
               1    STRONGLY AGREE
               2    AGREE
               3    DISAGREE
               4    STRONGLY DISAGREE
               8 M  DONT KNOW
               9 M  NO ANSWER

OPTIMIST (58) I AM ALWAYS OPTIMISTIC ABOUT MY FUTURE
           Measurement level: Ordinal
           Missing Values: 0, 8, 9
           Value    Label
               0 M  NAP
               1    STRONGLY AGREE
               2    AGREE
               3    DISAGREE
               4    STRONGLY DISAGREE
               8 M  DONT KNOW
               9 M  NO ANSWER
```

```
PESSIMST (59) I HARDLY EVER EXPECT THINGS TO GO MY WAY
           Measurement level: Ordinal
           Missing Values: 0, 8, 9
           Value    Label
               0 M  NAP
               1    STRONGLY AGREE
               2    AGREE
               3    DISAGREE
               4    STRONGLY DISAGREE
               8 M  DONT KNOW
               9 M  NO ANSWER

NOTCOUNT (60) I RARELY COUNT ON GOOD THINGS HAPPENING TO ME
           Measurement level: Ordinal
           Missing Values: 0, 8, 9
           Value    Label
               0 M  NAP
               1    STRONGLY AGREE
               2    AGREE
               3    DISAGREE
               4    STRONGLY DISAGREE
               8 M  DONT KNOW
               9 M  NO ANSWER

MOREGOOD (61) I EXPECT MORE GOOD THINGS TO HAPPEN TO ME THAN BAD
           Measurement level: Ordinal
           Missing Values: 0, 8, 9
           Value    Label
               0 M  NAP
               1    STRONGLY AGREE
               2    AGREE
               3    DISAGREE
               4    STRONGLY DISAGREE
               8 M  DONT KNOW
               9 M  NO ANSWER

EVDRINK (62) EVER DRANK ANY ALCOHOLIC BEVERAGE
           Measurement level: Ordinal
           Missing Values: 9
           Value    Label
               0    NO
               1    YES
```

```
DRINKYR (63) HOW REGULARLY R DRINK OVER LAST 12 MONTHS
           Measurement level: Ordinal
           Missing Values: 0, 98, 99
           Value    Label
               0 M  NAP
               1    Never in those 12 months
               2    1 to 3 times in 12 months
               3    4 to 7 times in 12 months
               4    8 to 11 times in 12 months
               5    1 to 3 times a month
               6    Once or twice a week
               7    3 to 4 times a week
               8    5 times a week or more
              98 M  DONT KNOW
              99 M  NO ANSWER

DRINKDAY (64) HOW MANY DRINKS R HAVE ON A DAY WHEN DRINKING
           Measurement level: Scale
           Missing Values: 0, 98, 99
           Value    Label
               0 M  NAP
              98 M  DONT KNOW
              99 M  NO ANSWER

OWNGUN (65) HAVE GUN IN HOME
           Measurement level: Nominal
           Missing Values: 3, 8, 9
           Value    Label
               0    NO
               1    YES

HUNT (66) DOES R OR SPOUSE HUNT
           Measurement level: Nominal
           Missing Values: 8, 9
           Value    Label
               1    RESP
               2    SPOUSE
               3    BOTH
               4    NEITHER
               8 M  DK
               9 M  NA

NEWS (67) HOW OFTEN DOES R READ NEWSPAPER
           Measurement level: Ordinal
           Missing Values: 8, 9
           Value    Label
               1    EVERYDAY
               2    FEW TIMES A WEEK
               3    ONCE A WEEK
               4    LESS THAN ONCE WK
               5    NEVER
               8 M  DK
               9 M  NA
```

```
TVHOURS (68) HOURS PER DAY WATCHING TV
          Measurement level: Scale
          Missing Values: 98, 99
          Value    Label
             -1    NAP
             98 M  DK
             99 M  NA

GETAHEAD (69) OPINION OF HOW PEOPLE GET AHEAD
          Measurement level: Nominal
          Missing Values: 8, 9
          Value    Label
              1    HARD WORK
              2    BOTH EQUALLY
              3    LUCK OR HELP
              8 M  DK
              9 M  NA

RACDIF1 (70) RACE DIFFERENCES DUE TO DISCRIMINATION
          Measurement level: Nominal
          Missing Values: 9
          Value    Label
              0    NO
              1    YES

RACDIF2 (71) RACE DIFFERENCES DUE TO INBORN DISABILITY
          Measurement level: Nominal
          Missing Values: 9
          Value    Label
              0    NO
              1    YES

RACDIF3 (72) RACE DIFFERENCES DUE TO LACK OF EDUCATION
          Measurement level: Nominal
          Missing Values: 9
          Value    Label
              0    NO
              1    YES

RACDIF4 (73) RACE DIFFERENCES DUE TO LACK OF WILL
          Measurement level: Nominal
          Missing Values: 9
          Value    Label
              0    NO
              1    YES

FEAR (74) AFRAID TO WALK AT NIGHT IN NEIGHBORHOOD
          Measurement level: Nominal
          Missing Values: 9
          Value    Label
              0    NO
              1    YES
```

```
EXCLDIMM (75) AMERICA SHOULD EXCLUDE ILLEGAL IMMIGRANTS
           Measurement level: Ordinal
           Missing Values: 8, 9
           Value    Label
               1    AGREE STRONGLY
               2    AGREE
               3    NEITHER AGREE NOR DISAGREE
               4    DISAGREE
               5    DISAGREE STRONGLY
               8 M  CANT CHOOSE
               9 M  NA

HGUNLAW (76) SHOULD BE MORE RESTRICTIONS ON HANDGUN
           Measurement level: Nominal
           Missing Values: 9
           Value    Label
               0    DISAGREE
               1    AGREE

PARTYID (77) POLITICAL PARTY AFFILIATION
           Measurement level: Ordinal
           Missing Values: 8, 9
           Value    Label
               0    STRONG DEMOCRAT
               1    NOT STR DEMOCRAT
               2    IND,NEAR DEM
               3    INDEPENDENT
               4    IND,NEAR REP
               5    NOT STR REPUBLICAN
               6    STRONG REPUBLICAN
               7    OTHER PARTY
               8 M  DK
               9 M  NA

VOTE00 (78) DID R VOTE IN 2000 ELECTION
           Measurement level: Nominal
           Missing Values: 4, 8, 9
           Value    Label
               1    VOTED
               2    DID NOT VOTE
               3    INELIGIBLE
               4 M  REFUSED TO ANSWER
               8 M  DONT KNOW/REMEMBER
               9 M  NA

PRES00 (79) VOTE FOR GORE, BUSH, NADER
           Measurement level: Nominal
           Missing Values: 8, 9
           Value    Label
               1    GORE
               2    BUSH
               3    NADER
               4    OTHER (SPECIFY)
               6    DIDNT VOTE
               8 M  DONT KNOW
               9 M  NA
```

```
POLVIEWS (80) THINK OF SELF AS LIBERAL OR CONSERVATIVE
          Measurement level: Ordinal
          Missing Values: 8, 9
          Value    Label
              1    EXTREMELY LIBERAL
              2    LIBERAL
              3    SLIGHTLY LIBERAL
              4    MODERATE
              5    SLGHTLY CONSERVATIVE
              6    CONSERVATIVE
              7    EXTRMLY CONSERVATIVE
              8 M  DK
              9 M  NA

NATENVIR (81) IMPROVING & PROTECTING ENVIRONMENT
          Measurement level: Ordinal
          Missing Values: 8, 9
          Value    Label
              1    TOO LITTLE
              2    ABOUT RIGHT
              3    TOO MUCH
              8 M  DK
              9 M  NA

NATHEAL (82) IMPROVING & PROTECTING NATIONS HEALTH
          Measurement level: Ordinal
          Missing Values: 8, 9
          Value    Label
              1    TOO LITTLE
              2    ABOUT RIGHT
              3    TOO MUCH
              8 M  DK
              9 M  NA

NATCRIME (83) HALTING RISING CRIME RATE
          Measurement level: Ordinal
          Missing Values: 8, 9
          Value    Label
              1    TOO LITTLE
              2    ABOUT RIGHT
              3    TOO MUCH
              8 M  DK
              9 M  NA

NATEDUC (84) IMPROVING NATIONS EDUCATION SYSTEM
          Measurement level: Ordinal
          Missing Values: 8, 9
          Value    Label
              1    TOO LITTLE
              2    ABOUT RIGHT
              3    TOO MUCH
              8 M  DK
              9 M  NA
```

```
[illegible]MS (85) MILITARY, ARMAMENTS, AND DEFENSE
           Measurement level: Ordinal
           Missing Values: 8, 9
           Value    Label
               1    TOO LITTLE
               2    ABOUT RIGHT
               3    TOO MUCH
               8 M  DK
               9 M  NA

NATAID (86) FOREIGN AID
           Measurement level: Ordinal
           Missing Values: 8, 9
           Value    Label
               1    TOO LITTLE
               2    ABOUT RIGHT
               3    TOO MUCH
               8 M  DK
               9 M  NA

NATFARE (87) WELFARE
           Measurement level: Ordinal
           Missing Values: 8, 9
           Value    Label
               1    TOO LITTLE
               2    ABOUT RIGHT
               3    TOO MUCH
               8 M  DK
               9 M  NA

NATSOC (88) SOCIAL SECURITY
           Measurement level: Ordinal
           Missing Values: 8, 9
           Value    Label
               1    TOO LITTLE
               2    ABOUT RIGHT
               3    TOO MUCH
               8 M  DK
               9 M  NA

NATCHLD (89) ASSISTANCE FOR CHILDCARE
           Measurement level: Ordinal
           Missing Values: 8, 9
           Value    Label
               1    TOO LITTLE
               2    ABOUT RIGHT
               3    TOO MUCH
               8 M  DK
               9 M  NA
```

```
NATSCI (90) SUPPORTING SCIENTIFIC RESEARCH
           Measurement level: Ordinal
           Missing Values: 8, 9
           Value    Label
               1    TOO LITTLE
               2    ABOUT RIGHT
               3    TOO MUCH
               8 M  DK
               9 M  NA

NATFAREY (91) ASSISTANCE TO THE POOR -- VERSION Y
           Measurement level: Ordinal
           Missing Values: 8, 9
           Value    Label
               1    TOO LITTLE
               2    ABOUT RIGHT
               3    TOO MUCH
               8 M  DK
               9 M  NA

CAPPUN (92) FAVOR OR OPPOSE DEATH PENALTY FOR MURDER
           Measurement level: Nominal
           Missing Values: 9
           Value    Label
               0    OPPOSE
               1    FAVOR

GUNLAW (93) FAVOR OR OPPOSE GUN PERMITS
           Measurement level: Nominal
           Missing Values: 9
           Value    Label
               0    OPPOSE
               1    FAVOR

COURTS (94) COURTS DEALING WITH CRIMINALS
           Measurement level: Ordinal
           Missing Values: 8, 9
           Value    Label
               1    TOO HARSH
               2    NOT HARSH ENOUGH
               3    ABOUT RIGHT
               8 M  DK
               9 M  NA

GRASS (95) SHOULD MARIJUANA BE MADE LEGAL
           Measurement level: Nominal
           Missing Values: 9
           Value    Label
               0    NOT LEGAL
               1    LEGAL
```

```
[illegible]MACT (96) FAVOR PREFERENCE IN HIRING BLACKS
          Measurement level: Ordinal
          Missing Values: 8, 9
          Value    Label
              1    STRONGLY SUPPORT PREF
              2    SUPPORT PREF
              3    OPPOSE PREF
              4    STRONGLY OPPOSE PREF
              8 M  DK
              9 M  NA

WRKWAYUP (97) BLACKS OVERCOME PREJUDICE WITHOUT FAVORS
          Measurement level: Ordinal
          Missing Values: 8, 9
          Value    Label
              1    AGREE STRONGLY
              2    AGREE SOMEWHAT
              3    NEITHER AGREE NOR DISAGREE
              4    DISAGREE SOMEWHAT
              5    DISAGREE STRONGLY
              8 M  DK
              9 M  NA

WORKWHTS (98) WHITES HARD WORKING - LAZY
          Measurement level: Ordinal
          Missing Values: 8, 9
          Value    Label
              1    HARDWORKING
              7    LAZY
              8 M  DONT KNOW
              9 M  NA

WORKBLKS (99) BLACKS HARD WORKING - LAZY
          Measurement level: Ordinal
          Missing Values: 8, 9
          Value    Label
              1    HARDWORKING
              7    LAZY
              8 M  DONT KNOW
              9 M  NA

INTLWHTS (100) WHITES UNINTELLIGENT -INTELLIGENT
          Measurement level: Ordinal
          Missing Values: 8, 9
          Value    Label
              1    UNINTELLIGENT
              7    INTELLIGENT
              8 M  DONT KNOW
              9 M  NA
```

```
INTLBLKS (101) BLACKS UNINTELLIGENT - INTELLIGENT
           Measurement level: Ordinal
           Missing Values: 8, 9
           Value     Label
               1     UNINTELLIGENT
               7     INTELLIGENT
               8 M   DONT KNOW
               9 M   NA

DISCAFF  102) WHITES HURT BY AFF. ACTION
           Measurement level: Ordinal
           Missing Values: 8, 9
           Value     Label
               1     VERY LIKELY
               2     SOMEWHAT LIKELY
               3     NOT VERY LIKELY
               8 M   DONT KNOW
               9 M   NA

FEJOBAFF (103) FOR OR AGAINST PREFERENTIAL HIRING OF WOMEN
           Measurement level: Ordinal
           Missing Values: 8, 9
           Value     Label
               1     STRONGLY FOR
               2     FOR
               3     AGAINST
               4     STRONGLY AGAINST
               8 M   DK
               9 M   NA

DISCAFFM (104) AFFIRM ACT. A MAN WON'T GET A JOB OR PROMOTION
           Measurement level: Ordinal
           Missing Values: 8, 9
           Value     Label
               1     VERY LIKELY
               2     SOMEWHAT LIKELY
               3     SOMEWHAT UNLIKELY
               4     VERY UNLIKELY
               8 M   DONT KNOW
               9 M   NA

DISCAFFW (105) AFFIRM ACT. A WOMAN WON'T GET A JOB OR PROMOTION
           Measurement level: Ordinal
           Missing Values: 8, 9
           Value     Label
               1     VERY LIKELY
               2     SOMEWHAT LIKELY
               3     SOMEWHAT UNLIKELY
               4     VERY UNLIKELY
               8 M   DONT KNOW
               9 M   NA
```

```
RELIG (106) RS RELIGIOUS PREFERENCE
          Measurement level: Nominal
          Missing Values: 98, 99
          Value    Label
              1    PROTESTANT
              2    CATHOLIC
              3    JEWISH
              4    NONE
              5    OTHER (SPECIFY)
              6    BUDDHISM
              7    HINDUISM
              8    OTHER EASTERN
              9    MOSLEM/ISLAM
             10    ORTHODOX-CHRISTIAN
             11    CHRISTIAN
             12    NATIVE AMERICAN
             13    INTER-NONDENOMINATIONAL
             98 M  DK
             99 M  NA

DENOM (107) SPECIFIC DENOMINATION
          Measurement level: Nominal
          Missing Values: 98, 99
          Value    Label
             10    AM BAPTIST ASSO
             11    AM BAPT CH IN USA
             12    NAT BAPT CONV OF AM
             13    NAT BAPT CONV USA
             14    SOUTHERN BAPTIST
             15    OTHER BAPTISTS
             18    BAPTIST-DK WHICH
             20    AFR METH EPISCOPAL
             21    AFR METH EP ZION
             22    UNITED METHODIST
             23    OTHER METHODIST
             28    METHODIST-DK WHICH
             30    AM LUTHERAN
             31    LUTH CH IN AMERICA
             32    LUTHERAN-MO SYNOD
             33    WI EVAN LUTH SYNOD
             34    OTHER LUTHERAN
             35    EVANGELICAL LUTH
             38    LUTHERAN-DK WHICH
             40    PRESBYTERIAN C IN US
             41    UNITED PRES CH IN US
             42    OTHER PRESBYTERIAN
             43    PRESBYTERIAN, MERGED
             48    PRESBYTERIAN-DK WH
             50    EPISCOPAL
             60    OTHER
             70    NO DENOMINATION
             98 M  DK
             99 M  NA
```

```
FUND (108) HOW FUNDAMENTALIST IS R CURRENTLY
           Measurement level: Ordinal
           Missing Values: 8, 9
           Value    Label
               1    FUNDAMENTALIST
               2    MODERATE
               3    LIBERAL
               8 M  DK
               9 M  NA-EXCLUDED

ATTEND (109) HOW OFTEN R ATTENDS RELIGIOUS SERVICES
           Measurement level: Ordinal
           Missing Values: 9
           Value    Label
               0    NEVER
               1    LT ONCE A YEAR
               2    ONCE A YEAR
               3    SEVRL TIMES A YR
               4    ONCE A MONTH
               5    2-3X A MONTH
               6    NRLY EVERY WEEK
               7    EVERY WEEK
               8    MORE THN ONCE WK
               9 M  DK,NA

RELITEN (110) STRENGTH OF AFFILIATION
           Measurement level: Ordinal
           Missing Values: 8, 9
           Value    Label
               1    STRONG
               2    NOT VERY STRONG
               3    SOMEWHAT STRONG
               4    NO RELIGION
               8 M  DK
               9 M  NA

POSTLIFE (111) BELIEF IN LIFE AFTER DEATH
           Measurement level: Nominal
           Missing Values: 9
           Value    Label
               0    NO
               1    YES

PRAY (112) HOW OFTEN DOES R PRAY
           Measurement level: Ordinal
           Missing Values: 8, 9
           Value    Label
               1    SEVERAL TIMES A DAY
               2    ONCE A DAY
               3    SEVERAL TIMES A WEEK
               4    ONCE A WEEK
               5    LT ONCE A WEEK
               6    NEVER
               8 M  DK
               9 M  NA
```

```
REBORN (113) HAS R EVER HAD A 'BORN AGAIN' EXPERIENCE
           Measurement level: Nominal
           Missing Values: 9
           Value    Label
               0    NO
               1    YES

BEAUSPRT (114) SPIRITUALLY TOUCHED BY BEAUTY CREATION
           Measurement level: Ordinal
           Missing Values: 8, 9
           Value    Label
               1    MANY TIMES A DAY
               2    EVERY DAY
               3    MOST DAYS
               4    SOME DAYS
               5    ONCE IN A WHILE
               6    NEVER ALMST NEVR
               8 M  DONT KNOW
               9 M  NA

RELEXP (115) HAVE RELIGIOUS EXPERIENCE CHANGED LIFE
           Measurement level: Nominal
           Missing Values: 9
           Value    Label
               0    NO
               1    YES

SEXEDUC (116) SEX EDUCATION IN PUBLIC SCHOOLS
           Measurement level: Nominal
           Missing Values: 8, 9
           Value    Label
               1    FAVOR
               2    OPPOSE
               3    DEPENDS
               8 M  DK
               9 M  NA

PREMARSX (117) SEX BEFORE MARRIAGE
           Measurement level: Ordinal
           Missing Values: 8, 9
           Value    Label
               1    ALWAYS WRONG
               2    ALMST ALWAYS WRG
               3    SOMETIMES WRONG
               4    NOT WRONG AT ALL
               5    OTHER
               8 M  DK
               9 M  NA
```

```
TEENSEX (118) SEX BEFORE MARRIAGE -- TEENS 14-16
          Measurement level: Ordinal
          Missing Values: 8, 9
          Value    Label
              1    ALWAYS WRONG
              2    ALMST ALWAYS WRG
              3    SOMETIMES WRONG
              4    NOT WRONG AT ALL
              5    OTHER
              8 M  DK
              9 M  NA

XMARSEX (119) SEX WITH PERSON OTHER THAN SPOUSE
          Measurement level: Ordinal
          Missing Values: 8, 9
          Value    Label
              1    ALWAYS WRONG
              2    ALMST ALWAYS WRG
              3    SOMETIMES WRONG
              4    NOT WRONG AT ALL
              5    OTHER
              8 M  DK
              9 M  NA

HOMOSEX (120) HOMOSEXUAL SEX RELATIONS
          Measurement level: Ordinal
          Missing Values: 8, 9
          Value    Label
              1    ALWAYS WRONG
              2    ALMST ALWAYS WRG
              3    SOMETIMES WRONG
              4    NOT WRONG AT ALL
              5    OTHER
              8 M  DK
              9 M  NA

PORNLAW (121) FEELINGS ABOUT PORNOGRAPHY LAWS
          Measurement level: Ordinal
          Missing Values: 8, 9
          Value    Label
              1    ILLEGAL TO ALL
              2    ILLEGAL UNDER 18
              3    LEGAL
              8 M  DK
              9 M  NA

XMOVIE (122) SEEN X-RATED MOVIE IN LAST YEAR
          Measurement level: Nominal
          Missing Values: 9
          Value    Label
              0    NO
              1    YES
```

```
PARTNERS (123) HOW MANY SEX PARTNERS R HAD IN LAST YEAR
           Measurement level: Ordinal
           Missing Values: -1, 99
           Value    Label
              -1 M  NAP
               0    NO PARTNERS
               1    1 PARTNER
               2    2 PARTNERS
               3    3 PARTNERS
               4    4 PARTNERS
               5    5-10 PARTNERS
               6    11-20 PARTNERS
               7    21-100 PARTNERS
               8    MORE THAN 100 PARTNERS
               9    1 OR MORE, DK #
              95    SEVERAL
              98    DK
              99 M  NA

SEXFREQ (124) FREQUENCY OF SEX DURING LAST YEAR
           Measurement level: Ordinal
           Missing Values: 8, 9
           Value    Label
               0    NOT AT ALL
               1    ONCE OR TWICE
               2    ONCE A MONTH
               3    2-3 TIMES A MONTH
               4    WEEKLY
               5    2-3 PER WEEK
               6    4+ PER WEEK
               8 M  DK
               9 M  NA

NUMWOMEN (125) NUMBER OF FEMALE SEX PARTNERS SINCE 18
           Measurement level: Scale
           Missing Values: -1, 999
           Value    Label
              -1 M  NAP
             999 M  NA

NUMMEN (126) NUMBER OF MALE SEX PARTNERS SINCE 18
           Measurement level: Scale
           Missing Values: -1, 999
           Value    Label
              -1 M  NAP
             999 M  NA
```

```
WKSTRESS (127) HOW OFTEN R FIND HER WORK STRESSFUL
           Measurement level: Ordinal
           Missing Values: 8, 9
           Value    Label
               1    ALWAYS
               2    OFTEN
               3    SOMETIMES
               4    HARDLY EVER
               5    NEVER
               8 M  DONT KNOW
               9 M  REFUSED

FECHLD (128) MOTHER WORKING DOESNT HURT CHILDREN
           Measurement level: Ordinal
           Missing Values: 8, 9
           Value    Label
               1    STRONGLY AGREE
               2    AGREE
               3    DISAGREE
               4    STRONGLY DISAGREE
               8 M  DK
               9 M  NA

FEPRESCH (129) PRESCHOOL KIDS SUFFER IF MOTHER WORKS
           Measurement level: Ordinal
           Missing Values: 8, 9
           Value    Label
               1    STRONGLY AGREE
               2    AGREE
               3    DISAGREE
               4    STRONGLY DISAGREE
               8 M  DK
               9 M  NA

FEFAM (130) BETTER FOR MAN TO WORK, WOMAN TEND HOME
           Measurement level: Ordinal
           Missing Values: 8, 9
           Value    Label
               1    STRONGLY AGREE
               2    AGREE
               3    DISAGREE
               4    STRONGLY DISAGREE
               8 M  DK
               9 M  NA

MEOVRWRK (131) MEN HURT FAMILY WHEN FOCUS ON WORK TOO MUCH
           Measurement level: Ordinal
           Missing Values: 8, 9
           Value    Label
               1    STRONGLY AGREE
               2    AGREE
               3    NEITHER AGREE NOR DISAGREE
               4    DISAGREE
               5    STRONGLY DISAGREE
               8 M  CAN T CHOOSE
               9 M  NA
```

```
WT2004 (132) WEIGHTS FOR 2004
           Measurement level: Scale

WT2004NR (133) WEIGHTS FOR 2004 AREA NONRESPONSE ADJUSTMENT
           Measurement level: Scale
```

APPENDIX 3: GSS04 QUESTION PHRASING

NAME	VARIABLE LABEL AND *QUESTION PHRASING*
YEAR	GSS YEAR FOR THIS RESPONDENT
ID	RESPONDNT ID NUMBER
MARITAL	MARITAL STATUS
	Are you currently married, widowed, divorced, separated, or have you never been married?
AGE	AGE OF RESPONDENT
	RESPONDENT'S AGE
EDUC	HIGHEST YEAR OF SCHOOL COMPLETED
	RESPONDENT'S EDUCATION
DEGREE	RS HIGHEST DEGREE
	RESPONDENT'S DEGREE
SEX	RESPONDENTS SEX
	CODE RESPONDENT'S SEX
RACE	RACE OF RESPONDENT
	What race do you consider yourself?
RACECEN1	WHAT IS RS RACE 1ST MENTION
	What is your race? Indicate one or more races that you consider yourself to be. First mention.
RACECEN2	WHAT IS RS RACE 2ND MENTION
	What is your race? Indicate one or more races that you consider yourself to be. Second mention.
RACECEN3	WHAT IS RS RACE 3RD MENTION
	What is your race? Indicate one or more races that you consider yourself to be. Third mention.
COHORT	YEAR OF BIRTH
	Birth cohort of respondent.
ZODIAC	RESPONDENTS ASTROLOGICAL SIGN
	ASTROLOGICAL SIGN OF RESPONDENT
DIVORCE	EVER BEEN DIVORCED OR SEPARATED
	IF CURRENTLY MARRIED OR WIDOWED: Have you ever been divorced or legally separated?
WIDOWED	EVER BEEN WIDOWED
	IF CURRENTLY MARRIED, SEPARATED, OR DIVORCED: Have you ever been widowed?
SIBS	NUMBER OF BROTHERS AND SISTERS
	How many brothers and sisters did you have? Please count those born alive, but no longer living, as well as those alive now. Also include stepbrothers and stepsisters, and children adopted by:
CHILDS	NUMBER OF CHILDREN
	How many children have you ever had? Please count all that were born alive at any time (including any you had from a previous marriage).
AGEKDBRN	R'S AGE WHEN 1ST CHILD BORN
	How old were you when your first child was born?
HOMPOP	NUMBER OF PERSONS IN HOUSEHOLD
	Household Size and Composition
BABIES	HOUSEHOLD MEMBERS LESS THAN 6 YRS OLD
	NUMBER OF MEMBERS UNDER 6. UNDER 6 YEARS:
PRETEEN	HOUSEHOLD MEMBERS 6 THRU 12 YRS OLD
	NUMBER OF MEMBERS 6 TO 12 YEARS. 6-12 YEARS:
TEENS	HOUSEHOLD MEMBERS 13 THRU 17 YRS OLD
	NUMBER OF MEMBERS 13 TO 17 YEARS OLD. 13-17 YEARS.
ADULTS	HOUSEHOLD MEMBERS 18 YRS AND OLDER
	NUMBER OF MEMBERS OVER I7 YEARS OLD. I8+ YEARS.
UNRELAT	NUMBER IN HOUSEHOLD NOT RELATED
	How many persons in the household are not related to you in any way?

RESPNUM	NUMBER IN FAMILY OF R
	Number in family of respondent.
HHTYPE1	HOUSEHOLD TYPE (CONDENSED)
	Household type (condensed).
PRESTG80	RS OCCUPATIONAL PRESTIGE SCORE (1980)
	PRESTIGE OF RESPONDENT'S OCCUPATION
SPPRES80	SPOUSES OCCUPATIONAL PRESTIGE SCORE (1980)
	PRESTIGE OF RESPONDENT'S SPOUSE'S OCCUPATION
INCOME98	TOTAL FAMILY INCOME
	In which of these groups did your total family income, from all sources, fall last year before taxes, that is?
RINCOME98	RESPONDENTS INCOME
	Did you earn any income from (OCCUPATION DESCRIBED IN Q. 2) in [1973/74/75/76/77/79/80/82/83/84/85/86/87/88/89/90/92/93/95/97/99/01/03]? IF YES: In which of these groups did your earnings from (OCCUPATION IN Q. 2) for last year[1973-2003]fall? That is, before taxes or other deductions.
EARNRS	HOW MANY IN FAMILY EARNED MONEY
	Just thinking about your family now those people in the household who are related to you . . . How many persons in the family, including yourself, earned any money last year- from any job or employment?
WRKSTAT	LABOR FRCE STATUS
	Last week were you working full time, part time, going to school, keeping house, or what?
HRS1	NUMBER OF HOURS WORKED LAST WEEK
	IF WORKING, FULL OR PART TIME: How many hours did you work last week, at all jobs?
WRKSLF	R SELF-EMP OR WORKS FOR SOMEBODY
	RESPONDENT'S EMPLOYMENT STATUS
PARTFULL	WAS R'S WORK PART-TIME OR FULL-TIME?
	When you worked in [I 993/95/97/99/2001/2003], was it usually full or part time?
ISCO88	RESPONDENT'S OCCUPATION, 1988 CENSUS
	Respondent's Occupation, I988 Census
SPISCO88	R'S SPOUSE'S OCCUPATION, 1988 CENSUS
	Respondent's Spouse's Occupation, I988 Census
SPWRKSTA	SPOUSE LABOR FORCE STATUS
	Last week was your (wife/husband) working full time, part time, going to school, keeping house, or what?
SPHRS1	NUMBER OF HRS SPOUSE WORKED LAST WEEK
	IF WORKING, FULL OR PART TIME: How many hours did (helshe) work last week, at all jobs?
SPEVWORK	SPOUSE EVER WORK AS LONG AS A YEAR
	IF RETIRED, IN SCHOOL, KEEPING HOUSE, OR OTHER: Did (he/she) ever work for as long as one year?
DWELOWN	DOES R OWN OR RENT HOME?
	(Do youlDoes your family) own your (homelapartment), pay rent, or what?
SEI	RESPONDENT SOCIOECONOMIC INDEX
	Respondent socioeconomic index.
SPSEI	R'S SPOUSE'S SOCIOECONOMIC INDEX
	Socioeconomic index of respondent's spouse.
HAPMAR	HAPPINESS OF MARRIAGE
	Taking things all together, how would you describe your marriage? Would you say that your marriage is very happy, pretty happy, or not too happy?
DIVLAW	DIVORCE LAWS
	Should divorce in this country be easier or more difficult to obtain than it is now?
SPANKING	FAVOR SPANKING TO DISCIPLINE CHILD
	Do you strongly agree, agree, disagree, or strongly disagree that it is sometimes necessary to discipline a child with a good, hard spanking?
MARHOMO	HOMOSEXUALS SHOULD HAVE RIGHT TO MARRY
	Homosexual couples should have the right to marry one another.

HAPPY	GENERAL HAPPINESS
	Taken all together, how would you say things are these dayswould you say that you are very happy, pretty happy, or not too happy?
HEALTH	CONDITION OF HEALTH
	Would you say your own health, in general, is excellent, good, fair, or poor?
LIFE	IS LIFE EXCITING OR DULL
	In general, do you find life exciting, pretty routine, or dull?
LETDIE1	ALLOW INCURABLE PATIENTS TO DIE
	Would you say that most of the time people try to be helpful, or that they are mostly just looking out for themselves?
RUSHED	HOW OFTEN R FEELS RUSHED
	In general, how do you feel about your timewould you say you always feel rushed even to do things you have to do, only sometimes feel rushed, or almost never feel rushed.
HEALTH1	RS HEALTH IN GENERAL
	Would you say that in general your health is Excellent, Very good, Good, Fair, or Poor?
MNTLHLTH	DAYS OF POOR MENTAL HEALTH PAST 30 DAYS
	Now thinking about your mental health, which includes stress, depression, and problems with emotions, for how many days during the past 30 days was your mental health not good?
SATSELF	ON THE WHOLE I AM SATISFIED WITH MYSELF
	Indicate your agreement with each of the following statements by selecting the number that comes closest to your answer: A. On the whole, I am satisfied with myself
AFAILURE	I AM INCLINED TO FEEL I AM A FAILURE
	All in all, I'm inclinded to feel I'm a failure.
SLFRSPCT	I WISH I COULD HAVE MORE RESPECT FOR MYSELF
	I wish I could have more respect for myself.
OFWORTH	I AM A PERSON OF WORTH AT LEAST EQUAL TO OTHERS
	I feel that I'm a person of worth, at least equal to others.
NOGOOD	AT TIMES I THINK I AM NO GOOD AT ALL
	At times I think I am no good at all.
OPTIMIST	I AM ALWAYS OPTIMISTIC ABOUT MY FUTURE
	I'm always optimistic about my future.
PESSIMST	I HARDLY EVER EXPECT THINGS TO GO MY WAY
	I hardly ever expect things to go my way.
NOTCOUNT	I RARELY COUNT ON GOOD THINGS HAPPENING TO ME
	I hardly ever expect things to go my way.
MOREGOOD	I EXPECT MORE GOOD THINGS TO HAPPEN TO ME THAN BAD
	Overall, I expect more good things to happen to me than bad.
EVDRINK	EVER DRANK ANY ALCOHOLIC BEVERAGE
	Was there ever a time when you drank any alcoholic beverages?
DRINKYR	HOW REGULARLY R DRINK OVER LAST 12 MONTHS
	Thinking back over the last 12 months, about how regularly did you drink alcoholic beverages? Would you say that it was…
DRINKDAY	HOW MANY DRINKS R HAVE ON A DAY WHEN DRINKING
	Again, as you think back over the last 12 months, about how many drinks would you have on a typical day when you drank? PROBE IF NECESSARY: Would it have been 5 or more?
OWNGUN	HAVE GUN IN HOME
	Do you happen to have in your home (IF HOUSE: or garage) any guns or revolvers?
HUNT	DOES R OR SPOUSE HUNT
	Do you (or does your [husbandlwife]) go hunting?
NEWS	HOW OFTEN DOES R READ NEWSPAPER
	How often do you read the newspaper every day, a few times a week, once a week, less than once a week, or never?
TVHOURS	HOURS PER DAY WATCHING TV

	On the average day, about how many hours do you personally watch television?
	OPINION OF HOW PEOPLE GET AHEAD
	Some people say that people get ahead by their own hard work; others say that lucky breaks or help from other people are more important. Which do you think is most important?
	RACE DIFFERENCES DUE TO DISCRIMINATION
	*On the average (NegroeslBlackslAfrican-Americans) have worse jobs, income, and housing than white people. Do you think these differences are:*Mainly due to discrimination?
RAC[illegible]	RACE DIFFERENCES DUE TO INBORN DISABILITY
	On the average (NegroeslBlackslAfrican-Americans) have worse jobs, income, and housing than white people. Do you think these differences are: Because most (NegroeslBlackslAfrican-Americans) have less inborn ability to learn?
RACDIF3	RACE DIFFERENCES DUE TO LACK OF EDUCATION
	On the average (NegroeslBlackslAfrican-Americans) have worse jobs, income, and housing than white people. Do you think these differences are: Because most (NegroeslBlackslAfrican-Americans) don't have the chance for education that it takes to rise out of poverty?
RACDIF4	RACE DIFFERENCES DUE TO LACK OF WILL
	On the average (NegroeslBlackslAfrican-Americans) have worse jobs, income, and housing than white people. Do you think these differences are: Because most (NegroeslBlackslAfrican-Americans) just don't have the motivation or will power to pull themselves up out of poverty?
FEAR	AFRAID TO WALK AT NIGHT IN NEIGHBORHOOD
	Is there any area right around here that is, within a mile, where you would be afraid to walk alone at night?
EXCLDIMM	AMERICA SHOULD EXCLUDE ILLEGAL IMMIGRANTS
	How much do you agree or disagree with the following statement?: America should take stronger measures to exclude illegal immigrants.
HGUNLAW	SHOULD BE MORE RESTRICTIONS ON HANDGUN
	Please tell me whether you agree or disagree with the following statement: "There should be more legal on handguns in our society."
PARTYID	POLITICAL PARTY AFFILIATION
	Is (NAME) a Democrat, a Republican, an Independent, or what? (Party Identification of First Person).
VOTE00	DID R VOTE IN 2000 ELECTION
	In 2000, you remember that Gore ran for President on the Democratic ticket against Bush for the Republicans . Do you remember for sure whether or not you voted in that election?
PRES00	VOTE FOR GORE, BUSH, NADER
	IF VOTED: Did you vote for Gore or Bush?
POLVIEWS	THINK OF SELF AS LIBERAL OR CONSERVATIVE
	We hear a lot of talk these days about liberals and conservatives. I'm going to show you a sevenpoint scale on which the political views that people might hold are arranged from extremely liberalpoint 1 t o extremely conservative--point 7. Where would you place yourself on this scale?
NATENVIR	IMPROVING & PROTECTING ENVIRONMENT
	We are faced with many problems in this country, none of which can be solved easily or inexpensively. I'm going to name some of these problems, and for each one I'd like you to tell me whether you think we're spending too much money on it, too little money, or about the right amount. First (READ ITEM A) . . . are we spending too much, too little, or about the right amount on (ITEM)? Improving and protecting the environment.
NATHEAL	IMPROVING & PROTECTING NATIONS HEALTH
	We are faced with many problems in this country, none of which can be solved easily or inexpensively. I'm going to name some of these problems, and for each one I'd like you to tell me whether you think we're spending too much money on it, too little money, or about the right amount. First (READ ITEM A) . . . are we spending too much, too little, or about the right amount on (ITEM)? Improving and protecting the nation's health.
NATCRIME	HALTING RISING CRIME RATE
	We are faced with many problems in this country, none of which can be solved easily or inexpensively. I'm going to name some of these problems, and for each one I'd like you to tell me whether you think we're spending too much money on it, too little money, or about the right amount. First (READ ITEM A) . . . are we spending too much, too little, or about the right amount on (ITEM)? Halting the rising crime rate.
NATEDUC	IMPROVING NATIONS EDUCATION SYSTEM

	We are faced with many problems in this country, none of which can be solved easily or inexpensively. I'm going to name some of these problems, and for each one I'd like you to tell me whether you think we're spending too much money on it, too little money, or about the right amount. First (READ ITEM A) . . . are we spending too much, too little, or about the right amount on (ITEM)? Improving the nation's education system.
NATARMS	MILITARY, ARMAMENTS, AND DEFENSE
	We are faced with many problems in this country, none of which can be solved easily or inexpensively. I'm going to name some of these problems, and for each one I'd like you to tell me whether you think we're spending too much money on it, too little money, or about the right amount. First (READ ITEM A) . . . are we spending too much, too little, or about the right amount on (ITEM)? The military, armaments and defense.
NATAID	FOREIGN AID
	We are faced with many problems in this country, none of which can be solved easily or inexpensively. I'm going to name some of these problems, and for each one I'd like you to tell me whether you think we're spending too much money on it, too little money, or about the right amount. First (READ ITEM A) . . . are we spending too much, too little, or about the right amount on (ITEM)? Foreign aid.
NATFARE	WELFARE
	We are faced with many problems in this country, none of which can be solved easily or inexpensively. I'm going to name some of these problems, and for each one I'd like you to tell me whether you think we're spending too much money on it, too little money, or about the right amount. First (READ ITEM A) . . . are we spending too much, too little, or about the right amount on (ITEM)? Welfare.
NATSOC	SOCIAL SECURITY
	We are faced with many problems in this country, none of which can be solved easily or inexpensively. I'm going to name some of these problems, and for each one I'd like you to tell me whether you think we're spending too much money on it, too little money, or about the right amount. First (READ ITEM A) . . . are we spending too much, too little, or about the right amount on (ITEM)? Social Security.
NATCHLD	ASSISTANCE FOR CHILDCARE
	We are faced with many problems in this country, none of which can be solved easily or inexpensively. I'm going to name some of these problems, and for each one I'd like you to tell me whether you think we're spending too much money on it, too little money, or about the right amount. First (READ ITEM A) . . . are we spending too much, too little, or about the right amount on (ITEM)? Assistance for childcare.
NATSCI	SUPPORTING SCIENTIFIC RESEARCH
	We are faced with many problems in this country, none of which can be solved easily or inexpensively. I'm going to name some of these problems, and for each one I'd like you to tell me whether you think we're spending too much money on it, too little money, or about the right amount. First (READ ITEM A) . . . are we spending too much, too little, or about the right amount on (ITEM)? Supporting scientific research.
NATFAREY	ASSISTANCE TO THE POOR -- VERSION Y
	We are faced with many problems in this country, none of which can be solved easily or inexpensively. I'm going to name some of these problems, and for each one I'd like you to tell me whether you think we're spending too much money on it, too little money, or about the right amount. First (READ ITEM A) . . . are we spending too much, too little, or about the right amount on (ITEM)? Assistance to the poor.
CAPPUN2	FAVOR OR OPPOSE DEATH PENALTY FOR MURDER
	Are you in favor of the death penalty for persons convicted of murder?
GUNLAW	FAVOR OR OPPOSE GUN PERMITS
	Would you favor or oppose a law which would require a person to obtain a police permit before he or she could buy a gun?
COURTS	COURTS DEALING WITH CRIMINALS
	In general, do you think the courts in this area deal too harshly or not harshly enough with criminals?
GRASS	SHOULD MARIJUANA BE MADE LEGAL
	Do you think the use of marijuana should be made legal or not?
AFFRMACT	FAVOR PREFERENCE IN HIRING BLACKS
	Some people say that because of past discrimination, blacks should be given preference in hiring and promotion. Others say that such preference in hiring and promotion of blacks is wrong because it discriminates against whites. What about your opinion -- are you for or against preferential hiring and promotion of blacks? IF FAVORS: A. Do you favor preference in hiring and promotion strongly or not strongly? IF OPPOSES: B. Do you oppose preference in hiring and promotion strongly or not strongly?
WRKWAYUP	BLACKS OVERCOME PREJUDICE WITHOUT FAVORS
	Do you agree strongly, agree somewhat, neither agree nor disagree, disagree somewhat, or disagree strongly with the following statement (HAND CARD TO RESPONDENT): Irish, Italians, Jewish and many other minorities overcame prejudice and worked their way up. Blacks should do the same without special favors.
WORKWHTS	WHITES HARD WORKING - LAZY

	The second set of characteristics asks if people in the group tend to be hard-working or if they tend to be lazy. Where would you rate whites in general on this scale?
WORKBLKS	BLACKS HARD WORKING - LAZY
	The second set of characteristics asks if people in the group tend to be hard-working or if they tend to be lazy. Where would you rate blacks in general on this scale?
INTLWHTS	WHITES UNINTELLIGENT -INTELLIGENT
	Do people in these groups tend to be unintelligent or tend to be intelligent? (WHITES)
INTLBLKS	BLACKS UNINTELLIGENT - INTELLIGENT
	Do people in these groups tend to be unintelligent or tend to be intelligent? (BLACKS)
DISCAFF	WHITES HURT BY AFF. ACTION
	What do you think the chances are these days that a white person won't get a job or promotion while an equally or less qualified black person gets one instead? Is this very likely, somewhat likely, or not very likely to happen these days?
FEJOBAFF	FOR OR AGAINST PREFERENTIAL HIRING OF WOMEN
	Some people say that because of past discrimination, women should be given preference in hiring and promotion. Others say that such preference in hiring and promotion of women is wrong because it discriminates against men. What about your opinion - are you for or against preferential hiring and promotion of women? IF FOR: Do you favor preference in hiring and promotion strongly or not strongly? IF AGAINST: Do you oppose preference in hiring and promotion strongly or not strongly?
DISCAFFM	AFFIRM ACT. A MAN WON'T GET A JOB OR PROMOTION
	What do you think the chances are these days that a man won't get a job or promotion while an equally or less qualified woman gets one instead? Is this very likely, somewhat likely, somewhat unlikely, or very unlikely these days?
DISCAFFW	AFFIRM ACT. A WOMAN WON'T GET A JOB OR PROMOTION
	What do you think the chances are these days that a woman won't get a job or promotion while an equally or less qualified man gets one instead? Is this very likely, somewhat likely, somewhat unlikely, or very unlikely these days?
RELIG	RS RELIGIOUS PREFERENCE
	What is your religious preference? Is it Protestant, Catholic, Jewish, some other religion, or no religion?
DENOM	SPECIFIC DENOMINATION
	What specific denomination is that, if any?
FUND	HOW FUNDAMENTALIST IS R CURRENTLY
	FundamentalismlLiberalism of Respondent's Religion
ATTEND	HOW OFTEN R ATTENDS RELIGIOUS SERVICES
	How often do you attend religious services?
RELITEN	STRENGTH OF AFFILIATION
	Would you call yourself a strong (PREFERENCE NAMED IN Q. 104 OR 104-A) or a not very strong (PREFERENCE NAMED IN Q. 104 OR 104-A (religious preference))?
POSTLIFE	BELIEF IN LIFE AFTER DEATH
	Do you believe there is a life after death?
PRAY	HOW OFTEN DOES R PRAY
	About how often do you pray?
REBORN	HAS R EVER HAD A 'BORN AGAIN' EXPERIENCE
	Would you say you have been "born again" or have had a "born again" experience - - that is, a turning point in your life when you committed yourself to Christ?
BEAUSPRT	SPIRITUALLY TOUCHED BY BEAUTY CREATION
	I am spiritually touched by the beauty of creation.
RELEXP	HAVE RELIGIOUS EXPERIENCE CHANGED LIFE
	Did you ever have a religious or spiritual experience that changed your life?
SEXEDUC	SEX EDUCATION IN PUBLIC SCHOOLS
	Would you be for or against sex education in the public schools?
PREMARSX	SEX BEFORE MARRIAGE
	There's been a lot of discussion about the way morals and attitudes about sex are changing in this country. If a man and woman have sex relations before marriage, do you think it is always wrong, almost always wrong, wrong only sometimes, or not wrong at all?
TEENSEX	SEX BEFORE MARRIAGE -- TEENS 14-16

	What if they are in their early teens, say 14 to 16 years old? In that case, do you think sex relati marriage are always wrong, almost always wrong, wrong only sometimes, or not wrong at all?
XMARSEX	SEX WITH PERSON OTHER THAN SPOUSE
	What about a married person having sexual relations with someone other than his or her husband . .
HOMOSEX	HOMOSEXUAL SEX RELATIONS
	What about sexual relations between two adults of the same s e x d o you think it is always wrong, al always wrong, wrong only sometimes, or not wrong at all?
PORNLAW	FEELINGS ABOUT PORNOGRAPHY LAWS
	Which of these statements comes closest to your feelings about pornography laws?
XMOVIE	SEEN X-RATED MOVIE IN LAST YEAR
	Have you seen an X-rated movie either in a movie theater or on a VCR (Video Cassette Recorder) in the last year?
PARTNERS	HOW MANY SEX PARTNERS R HAD IN LAST YEAR
	How many sex partners have you had in the last 12 months?
SEXFREQ	FREQUENCY OF SEX DURING LAST YEAR
	How many sex partners have you had in the last 12 months?
NUMWOMEN	NUMBER OF FEMALE SEX PARTNERS SINCE 18
	Now thinking about the time since your 18th birthday (including the past 12 months) how many female partners have you had sex with?
NUMMEN	NUMBER OF MALE SEX PARTNERS SINCE 18
	Now thinking about the time since your 18th birthday (including the past 12 months) how many male partners have you had sex with?
WKSTRESS	HOW OFTEN R FIND HER WORK STRESSFUL
	How often do you find your work stressful?
FECHLD	MOTHER WORKING DOESNT HURT CHILDREN
	A working mother can establish just as warm and secure a relationship with her children as a mother who does not work.
FEPRESCH	PRESCHOOL KIDS SUFFER IF MOTHER WORKS
	A preschool child is likely to suffer if his or her mother works
FEFAM	BETTER FOR MAN TO WORK, WOMAN TEND HOME
	It is much better for everyone involved if the man is the achiever outside the home and the woman takes care of the home and family.
MEOVRWRK	MEN HURT FAMILY WHEN FOCUS ON WORK TOO MUCH
	Family life often suffers because men concentrate too much on their work
WT2004	WEIGHTS FOR 2004
	Weight of variables.
WT2004NR	WEIGHTS FOR 2004 AREA NONRESPONSE ADJUSTMENT

APPENDIX 4: VARIABLE LABEL ABBREVIATIONS

Abbreviation	Text
DK	Don’t Know
Elem	Elementary
Second.	Secondary
Govt	Government
NA	No answer
NAP	Not applicable
Pop	Population

Permissions

Variables in the STATES07.SAV data are used by permission from Kathleen O'Leary Morgan and Scott Morgan (2007) *State Rankings 2007*, *Crime State Rankings 2007, and Health Care State Rankings 2007*, Morgan Quitno, Lawrence, Kansas.

Variables in the GSS04.SAV data are used by permission from the *Roper Center for Public Opinion Research*, Storrs, Ct. and the National Opinion Research Center at the University of Chicago, Chicago, IL.

References and Further Reading

Preface

Howery, Carla and Havidan Rodriguez. 2006. "Integrating Data Analysis: Working with Sociology Departments to Address the Scientific Literacy Gap." *Teaching Sociology* 34:5-22.

Madison, Bernard and Lynn Arthur Steen. 2003. *Quantitative Literacy: Why Numeracy Matters for Schools and Colleges*. Princeton, NJ: National Council on Education and the Disciplines.

McKinney, Kathleen, Carla Howery, Kerry Strand, Edward Kain, and Catherine White Berheide. 2004. "Liberal Learning and the Sociology Major Updated: Meeting the Challenge of Teaching Sociology in the Twenty-First Century." American Sociological Association, Washington DC.

Shaeffer, Richard. 2003. "Statistics and Quantitative Literacy." Pp. 145-152 in *Quantitative Literacy: Why Numeracy Matters for Schools and Colleges*, edited by Bernard Madison and Lynn Arthur Steen. Princeton, NJ: National Council on Education and the Disciplines.

Sweet, Stephen and Kerry Strand. 2006. "Cultivating Quantitative Literacy: The Role of Sociology." *Teaching Sociology* 34:1-4.

Universities, Association of American Colleges &. 2002. "Greater Expectations: A New Vision for Learning as the Nation Goes to College." Association of American Colleges & Universities, Washington, DC.

Chapter 1

Glaser, Barney and Anselm Straus. 1967. *The Discovery of Grounded Theory*. Chicago: Aldine.

Hoaglin, David, Frederick Mosteller, and John W. Tukey. 1983. *Understanding Robust and Exploratory Data Analysis*. New York: Wiley.

Kuhn, Thomas. 1997. *The Copernican Revolution: Planetary Astronomy in the Development of Western Thought*. New York, NY: MJF Books.

Neider, Charles. 2000 [1917]. "Autobiography of Mark Twain." New York: Harpers Perennial.

Sweet, Stephen and Peter Meiksins. 2008. *Changing Contours of Work: Jobs and Opportunities in the New Economy*. Thousand Oaks, CA: Pine Forge Press.

Data and Story Library: http://lib.stat.cmu.edu/DASL/

StatLib Data Set Archive: http://lib.stat.cmu.edu/datasets/

National Center for Education Statistics: http://nces.ed.gov/

Bureau of Justice Statistics: http://www.ojp.usdoj.gov/bjs/

Bureau of Labor Statistics: http://www.bls.gov/

Chapter 2

Morgan, Kathleen O'Leary, Scott Morgan, and Neal Quitno. 2007. *State Rankings 2000.* Lawrence, Kansas: Morgan Quitno.

Morgan, Kathleen O'Leary, Scott Morgan, and Neal Quitno. 2007. *Crime State Rankings 2000.* Lawrence, Kansas: Morgan Quitno.

Morgan, Kathleen O'Leary, Scott Morgan, and Neal Quitno. 2007. *Health Care State Rankings. 2000.* Lawrence, Kansas: Morgan Quitno

Chapter 4

Boslaugh, S. 2004. *An Intermediate Guide to SPSS Programming: Using Syntax for Data Management.* Thousand Oaks, CA: Sage Publications.

Chapter 5

Tufte, Edward. 1986. *The Visual Display of Quantitative Information*. Cheshire, Connecticut: Graphics Press.

Chapter 6

Keppel, Geoffrey and Thomas Wickens. 2007. *Design and Analysis, 5th Edition*. Upper Saddle River, NJ: Prentice Hall.

Montgomery, Douglas. 2005. *Design and Analysis of Experiments.* New York: John Wiley & Sons.

Kutner, Michael, John Neter, Christopher J. Nachtsheim, and William Wasserman. 2004. *Applied Linear Statistical Models*. Columbus, OH: McGraw-Hill.

Chapter 7

Berry, William. 1993. *Understanding Regression Assumptions*. Thousand Oaks, CA: Sage Publications.

Hardy, Melissa. 1993. *Regression with Dummy Variables*. Thousand Oaks, CA: Sage Publications.

Kutner, Michael, John Neter, Christopher J. Nachtsheim, and William Wasserman. 2004. *Applied Linear Statistical Models*. Columbus, OH: McGraw-Hill.

Chapter 8

Aldrich, John and Forrest Nelson. 1984. *Linear Probability, Logit and Probit Models*. Beverly Hills: Sage Publications.

Hosmer, David and Stanley Lemeshow. 2000. *Applied Logistic Regression*. New York: John Wiley and Sons.

Straus, Murray and Stephen Sweet. 1992. "Verbal/Symbolic Aggression in Couples: Incidence Rates and Relationships to Personal Characteristics." *Journal of Marriage and the Family* 54:346-357.

Chapter 9

American Psychological Association. 2001. *Publication Manual of the American Psychological Association*. 5th ed. Washington, DC: American Psychological Association.

American Sociological Association. 2007. *American Sociological Association Style Guide*. 3rd Edition. Washington DC: American Sociological Association.

Becker, Howard and Pamela Richards. 2007. *Writing for the Social Sciences: How to Start and Finish Your Thesis, Book, or Article, 2nd Edition*. Chicago, IL: University of Chicago Press.

Bem, Darryl. 2003. "Writing the Empirical Journal Article"._ in *The Compleat Academic: A Practical Guide for the Beginning Social Scientist, 2nd Edition*, edited by Darley, J. M., M. P. Zanna, & H. L. Roediger III. Washington, DC: American Psychological Association. Available at: http://dbem.ws/WritingArticle.pdf

Campbell, Donald and Julian Stanley. 1963. *Experimental and Quasi-Experimental Designs for Research*. Chicago: Rand McNally.

Index

NOTES